Architects
of Eternity

Architects
of Eternity

The new science of fossils

Richard Corfield

HEADLINE

First published in 2001
by HEADLINE BOOK PUBLISHING

10 9 8 7 6 5 4 3 2 1

British Library Cataloguing in Publication Data

Corfield, Richard
Architects of eternity
1. Paleontology
I. Title
560

ISBN 0 7472 7179 8 (hardback)
ISBN 0 7472 71801 (trade paperback)

Every effort has been made to fulfil requirements with regard to
reproducing copyright material. The author and publisher will be
glad to rectify any omissions at the earliest opportunity.

Typeset by Avon Dataset Ltd, Bidford-on-Avon, Warks

Text design by Tony Cohen

Printed and bound in Great Britain by
Mackays of Chatham Plc, Chatham, Kent

HEADLINE BOOK PUBLISHING
A division of Hodder Headline
338 Euston Road
London NW1 3BH

www.headline.co.uk
www.hodderheadline.com

For Julie
With my love

Acknowledgements

A book like this is an undertaking that relies fundamentally on the skill, kindness and generosity of many people. I am especially grateful to the friends and colleagues in Oxford and beyond who have taken the time and trouble to talk to me about the new science of fossils and listen to my ideas about the probable future of palaeontology. Several have also read portions of the manuscript. Thanks therefore to Steve Simpson, Stephen Moorbath, Santo Bains, Alan Cooper, Michael Durkin, Roy Goodwin, Owen Green, Mike Hall, Mary Morse, John O'Sullivan, Birger Schmitz and Derek Siveter. Thanks also to other friends: Gordon and Diane Evans, Paul Edgington and the Brill Boys at the Nut Tree, Murcott and Jon and Jo Oldham, Bob Cox, Peter James, John George, Marvina Houghton at the Royal Oak, Ramsden as well as Rob and Rachel Hancock. Thanks also to Dennis and Christine Armstrong and the fine members of the Frewen Club for support and encouragement.

A book about the history of palaeontology could not be written without the best libraries in the world: Oxford libraries are just that, so thanks to the librarians – most particularly at the Radcliffe Science Library – for their patience and good humour when I needed, at the shortest possible notice, an obscure text lurking in the deepest corner of the stack. Thanks also to Karen Smith who was invaluable in helping me track down material. Thanks too to Paul Smith and Jon Clatworthy in the Lapworth Archives at Birmingham University.

I am extremely grateful to my agent Peter Robinson of the Curtis Brown Group for perceptive advice. Thanks also to the wonderful team at Hodder Headline who have laboured tirelessly behind the scenes and made me feel so much a member of the Headline family: my editor Doug Young, Heather Holden-Brown and Jo Roberts-Miller. Thanks also to my copy-editor Jane Butcher for a careful and sensitive job on a manuscript that more than once lapsed into jargon.

Finally thanks to Julie, Jess and little Suzy for putting up with my furrowed brow in the evenings and weekends of the past year and a half. You have put up with Dad in the back room for far too long. I truly, truly couldn't have done it without you.

Contents

'The primary purpose of a liberal education is to make one's mind a pleasant place in which to spend one's time'

Thomas Henry Huxley, 1825–1895

CHAPTER ONE

The Size of Things to Come

What do you think about when you think of palaeontology? Is it dusty museum cabinets filled with the skeletal remains of long-dead fish gathering dust (tended perhaps by ageing professors who look rather like that curator vanishing into the distance as the credits roll at the end of *Raiders of the Lost Ark*)? Or is it perhaps ravening dinosaurs reconstituted from mended DNA chasing Jeff Goldblum along a remote jungle road in *Jurassic Park*?

The reason that I frame the question in this exaggerated way is that these two film clips represent two poles of the way that palaeontology – the science of fossils – is viewed. Not so many years ago the first stereotype would have been an accurate reflection of most people's perceptions, and yet now it is clear that many people, especially children (and their parents!), view fossils as living organisms that merely – and somewhat inconveniently – happen to be dead. The emphasis has shifted. Fossils are no longer dusty museum specimens: they are dead animals and plants in the process of resurrection by renewed public interest. Our perceptions have been indelibly and irremediably retouched by Michael Crichton and Steven Spielberg. We live now in the age of the media fossil.

So which then of these two perceptions is correct? Frankly, neither. Rather, they are outliers of a spectrum of scientific endeavour that grades seamlessly from the factual description of fossils (taxonomy) to the reconstruction of the diet of dinosaurs using tiny variations in the weight of their atomic remains (biogeochemistry).

It is this spectrum – this difference between the old and the new palaeontology – that is the subject of this book.

Why should we be interested in a subject that is so diffuse it is smeared across subject areas from descriptive biology to nuclear physics? Because it is this plurality and diversity that makes palaeontology unique and therefore among the most interesting of the sciences today. Taxonomy for its own sake is perceived, with some justification, as an outmoded description-based area that requires little in the way of hypothesis and scientific speculation; biogeochemistry on the other hand seems closer to science fiction than science fact. The simultaneous coexistence of the two within the frontiers of one subject is the frisson that made Michael Crichton's book so exciting and so successful.

I have spent the last fifteen years working in one of the areas that is close to the cutting edge of palaeontology. On a good day I even flatter myself that it is actually at the cutting edge! During these years I have been exposed to ideas and techniques that would quite certainly have been beyond the dreams of palaeontologists even fifty years ago. Collectively these comprise what I call the new science of fossils. It is this new palaeontology which depends for its success on the very latest in technology. Scanning electron microscopes with secondary X-ray detectors, mass spectrometers, DNA sequencers, the very latest in supercomputers are just some of the machines that are routinely used these days. Yet it is equally true that the new science of fossils can only be constructed on the foundations of the old – description and taxonomy of long-dead organisms are essential prerequisites.

To make this transition from the old to the new, palaeontology has borrowed heavily from other more mature disciplines. The classical sciences of my A-level youth, for example, physics, chemistry and biology, have provided the ideas and technologies that palaeontology has used to make itself into a subject that is much more than the sum of its parts. Yet this juxtaposition of the old and the new palaeontology is not an easy one; I call it the schizophrenic science. For within its borders there are still those whose world view of the science of fossils is merely the description of new – or sometimes even the re-description of previously described! – fossils. And yet in the same university, sometimes separated by not more than a few yards, there are those who are dissolving fossils into their component molecules and atoms and using these most elemental of remains to reconstruct the temperature and carbon dioxide content of ancient atmospheres and oceans. These are the scientists who are using fossils to inform us what will happen as mankind continues to pump gigatons of carbon dioxide into the atmosphere. The new science of fossils is about scientists who have stopped looking at fossils for their own sake – the 'Cor, look at that!' approach – and are starting to use fossils as vehicles that, by yielding the secrets of the past, actually inform our future. Palaeontology is studied in departments other than earth science departments, though. In biology departments across Europe, Australasia and North America particularly there are scientists (I call them 'reluctant palaeontologists') who, using recently developed and marvellously subtle techniques, are able to divine the relationships and even the elapsed time since evolutionary divergence between disparate groups of organisms. The reluctant palaeontologists reconstruct the history of life by analysing the differences between *living* genetic systems and their biochemical products. There are even reluctant palaeontologists who can *reconstruct* the genetic code of true fossils such as our own relatives, the Neanderthals.

'Science is all about people', my old professor at Cambridge once said to me. I have never forgotten that and the years that I have spent as a professional scientist have proved him right time and time again. Science is a very human drama of ambition, hard work, setback, jealousy, success and heartbreaking failure. To strip it of its personalities is to deny it its humanity. To simply describe the technical and scientific breakthroughs that have transformed the old science of fossils into the new is not enough. The story of the transformation is the story of those people who dared to look deeper and more carefully through shifting veils of time and separate truth from falsehood; it is the story of those who developed ideas and techniques that they and their students went on to exploit with devastating effect. The scientists who wrought these changes are heroes (and villains) at least as compelling as anything in fiction.

They are architects of eternity.

So I approach the new science of fossils from the point of view of telling a story. I also eschew the recent trend of portraying palaeontology as a branch of natural history, for I must here confess a personal bias; I am an unrepentant techie, a man who ever since he was a small boy has been in love with expensive machines (and the more flashing lights they had the better!). I was not the nice little boy whose eyes went round with wonder as the splitting of a Welsh slate exposed a fabulous trilobite to the first light of the sun in 400 million years; I was the horrible little boy who mixed sugar and sodium compounds in his back garden trying to recreate the blast radius at Bikini Atoll.

It was doubtless for this reason that I found myself drawn to the more high-tech aspects of palaeontology (specifically those used in the subdiscipline known as palaeoceanography, the study of ancient oceans) as a graduate student in Cambridge in the 1980s. But even for a technophilic palaeontologist like me there is still nowhere better to ply his trade than the city and university of Oxford where I have been since 1988. Oxford has a long and fine tradition in science generally and in palaeontology particularly and as you walk around the city and colleges the sense of history presses in upon you like a tangible entity. Oxford – mostly by virtue of its incredible natural history museum – was in on the very beginnings of the new science of fossils.

A word first about terminology. Palaeontology is hedged about with some of the most fearsome nouns in science. I'm quite sure that they've been enough to put many a prospective student off. But fear not:

Camels Ordinarily Sit Down Carefully, Perhaps Their Joints Creak? Possibly Early Oiling May Prevent Permanent Resting.

Got that? It's very simple and highly informative (see opposite for why).

As you can see, the names of the eras and periods of geological time on their own are pretty terrifying; this is simply my way of keeping them in their proper perspective and reminding me what they are – merely necessary

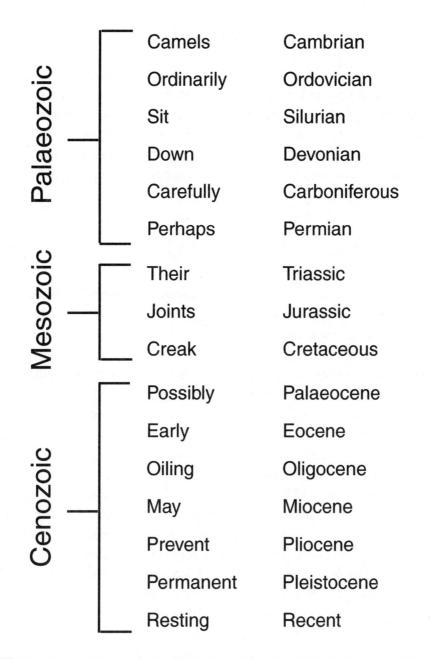

Palaeozoic	Camels	Cambrian
	Ordinarily	Ordovician
	Sit	Silurian
	Down	Devonian
	Carefully	Carboniferous
	Perhaps	Permian
Mesozoic	Their	Triassic
	Joints	Jurassic
	Creak	Cretaceous
Cenozoic	Possibly	Palaeocene
	Early	Eocene
	Oiling	Oligocene
	May	Miocene
	Prevent	Pliocene
	Permanent	Pleistocene
	Resting	Recent

Table 1.1 A useful way of remembering the divisions of Phanerozoic time. The dates of these series boundaries (as well as those of earlier intervals) are shown in the appendix.

names (many, if not most, with nineteenth-century origins) to help distinguish chunks of relative time. A word of warning though, the Camel mnemonic merely serves to distinguish the most commonly used divisions of geological time. Delve deeper and you will find that there are subdivisions whose names are very arcane indeed (and still subject to change; as we shall see, palaeontologists love to squabble over the names of sub-eras and the positions of period boundaries). To keep you orientated I've put a version of the geological timescale in an appendix and it includes these subdivisions.

The prehistory of palaeontology

I have chosen to start this narrative in Oxford in the mid-nineteenth century because it is a convenient time and place from which to trace the development of the new science of fossils. The roots of the new palaeontology can just be seen appearing at about this time. But I shall not dwell too long on them, because the crystallisation of the new science of fossils did not start in earnest until the end of the nineteenth century. By 1900 the infrastructure of the new palaeontology was in place and growing fast. This was the era of America's technological and scientific revolution, the time when the Rockefellers and Carnegies decided to invest in knowledge and science. So I shall only mention in passing those architects of what we may call the prehistory of palaeontology.

The science of palaeontology originated around the fifth century BC when the Greek philosophers Anaximander, Pythagoras and Herodotus proposed that the mineralised remains of lithified organisms must have been deposited in water. But this remarkably prescient observation was lost when the Greek civilisation fell and it was not until the fifteenth century that Leonardo da Vinci suggested again that fossil marine shells need not have originated where they were found but could have been transported by running water. In the seventeenth century, the Italian philosopher Lucilio Vanini even went so far as to suggest that humans might have evolved from apes. In the age of the Inquisition this was not a smart suggestion to make and Vanini was burned alive for the idea in 1619. Only thirty years later,

Archbishop Ussher made his famous pronouncement on the date of creation, based on the chronology of the Biblical prophets. This led to the calculation that the date the Earth was created could be narrowed down to 26 October 4004 BC; a date that we now know underestimates the true age of the Earth by six orders of magnitude. All in all the seventeenth century was characterised by a particularly unhealthy conflation of theology and natural history (a partnership that continued until well into the nineteenth century). In 1735 Linnaeus published his fabled *Systema Naturae* in which he proposed the adoption of the binomial classification system of organisms that is still in use today. Later in the seventeenth century, Abraham Gottlob Werner proposed the Neptunian theory, which asserted that all rocks were deposited by a primordial ocean. It was not until 1795 that James Hutton overturned this Neptunian theory of rock formation in his elaborate treatise *Theory of the Earth*, in which he suggested that the processes of rock creation were approximately balanced by processes of rock destruction. (Incidentally anticipating by 150 years the processes of rock formation and subduction at oceanic ridges and margins that are the centrepiece of plate tectonic theory today.)

The first of the great Oxford palaeontologists was William Smith who was born in the small Cotswold village of Churchill, not far from Moreton-in-Marsh in 1769. His beginnings were humble (he was the son of a blacksmith) but he ended his life as a man who would go down in history as the father of British geology. At the age of eighteen William Smith was employed as a land surveyor at the local town of Stow-on-the-Wold. He excelled at the work and these successes soon led to further employment. Before long he was asked to give evidence before Parliament in connection with the building of canals near the Somerset coalfields. His studies in pursuit of this commission consolidated his theories concerning the nature of rock layering – the so-called superposition of strata – and by the end of the eighteenth century he had concluded that rock layers could be readily traced across wide areas. He began to use colour on his maps to distinguish different beds as they outcropped (the formal term for the regions where rock layers intersect the surface of the land) across southern England. In 1799 Smith exhibited the very first geological map: the area around the city of Bath. The geological map remains today the most basic tool of the geologist and palaeontologist. Smith completed several geological maps of England and Wales and in 1815 he was able to complete a synthesis map entitled 'A delineation of the strata of England and Wales and part of Scotland'. By 1819, he had published his most important work, *Strata as identified by fossils*, the foundation of the palaeontological subdiscipline known as biostratigraphy, a subject that we shall examine in detail in Chapter 2.

If you visit Oxford's University Museum of Natural History today you will find in a corner of an upper gallery Smith's geological map of Oxfordshire (published in 1820). Examine it carefully and you will see that the county is bisected by a cream band indicating the extent of the outcrop of a rock layer known as the Cornbrash and towards its eastern extremity, on the margin of the drained marsh known as Ot Moor, you will see the tiny village of Islip. Drive up the shallow hill from Ot Moor and straight ahead is an ancient church. Buried in the churchyard there is William Buckland, like Smith, another of the great Oxford palaeontologists. He was born in Axminster in Devon in 1784 where his father was a clergyman. Buckland arrived in Oxford in 1801 to read divinity, just two years after Smith had exhibited the first geological map. Buckland took holy orders and was eventually elected a Fellow of Corpus in 1809. Finding the Oxford atmosphere convivial, he stayed on in the area and in 1813 was elected Reader in Mineralogy. By 1818 he had been elected to the newly established Readership in Geology. Buckland was an enormously energetic collector of fossils and minerals and his collections eventually came to form the core of Oxford's Museum of Natural History. His scientific interests were broad, too, for one of his most famous finds was a specimen of archaeological rather than palaeontological antiquity, the Cro Magnon skeleton from Wales known as the Red Lady of Paviland, which is still on display today in what was the 'University Museum', now renamed the University Museum of Natural History. In 1824 he wrote an account for the *Transactions of the Geological Society of London* of a giant skeleton that had been found in the small Cotswold village of Stonesfield, only five miles from where I write this. Buckland's contribution was the first published account of a dinosaur in the scientific literature. The skeleton was of *Megalosaurus*, a giant plant-eating dinosaur from the Jurassic period that walked erect on two legs. Buckland was also known for his interest in 'catastrophism', a doctrine that acknowledged that events in Earth's history could occur very rapidly and did not necessarily have a parallel in the present day. This principle was later superseded by the uniformitarian ideas of Charles Lyell. The uniformitarian doctrine holds that 'the present is the key to the past' and that events that have occurred over geological time are merely the same processes that happen in the present day but which have operated over vastly expanded timescales. In fact, it was only in the last two decades of the twentieth century that the idea of unique – or at least rare – events in the Earth's history came to be resurrected primarily through the work of the Alvarez father and son team working on the Cretaceous–Tertiary boundary, as we shall see later. Doubtless, Buckland would have relished this victory of his ideas.

Early in his career Buckland believed that the fossil bones, strange perched boulders known as glacial erratics and indeed just the general geography of Oxfordshire were all evidence for the Biblical flood, but in later years he came to reject this idea, becoming instead one of the earliest proponents of the new theory of glaciation. In an age when the difference between the clergy and academia was almost non-existent it was not surprising that Buckland continued to hold religious office. He was appointed Canon of Christ Church, Oxford in 1825 but in 1845, while still retaining his Oxford Readership, was appointed Dean of Westminster by the Prime Minister, Robert Peel. Despite Islip's proximity to the University it was not through geology that Buckland came to this quiet corner of Oxfordshire. The Dean of Westminster has charge of two churches, Westminster Abbey itself and also the country church at Islip where the job provided a country home. It was to here that Buckland retired and where he chose to be buried, in 1856.

William Buckland's successor at Oxford was John Phillips. Phillips, born in Marsland, Wiltshire in 1800, was the nephew of William Smith who brought him up and so it was that he learned his geology literally at the master's knee. He had a varied career, was curator of the York Museum, one of the founders of the British Association, Professor at King's College, London, Professor of Geology at Trinity College, Dublin but eventually wound up in Oxford in 1853 as a deputy Reader in Geology to William Buckland. After Buckland's death, Phillips became Reader in Geology and subsequently Professor. As well as teaching geology, Phillips played a leading role in the building of the University Museum and was its first keeper until his death in 1874. Among Phillips' research achievements was the introduction in 1841 of the concept of Palaeozoic, Mesozoic and Cenozoic, the fundamental divisions of palaeontological time. Although we now know that complex multicellular fossils existed in the late Precambrian – before the start of the Palaeozoic (*see* appendix) – Phillips' achievement as the first to codify the passage of relative time qualifies him as perhaps the first true architect of eternity. In 1852 Phillips was the first scientist to apply his geological expertise to the study of the moon using the great telescope which belonged to the Earl of Rosse. By 1853 he was recording his results photographically on collodium plates. In 1868 he began to draw analogies between many features of the Earth and the moon and had also started to publish on the structure of the Martian surface, an occupation that has strong resonance with the present day as we currently contemplate the possibility of life – at least at some point in its long history – on Mars. Phillips died in 1874.

But it was within Phillips' most tangible physical achievement – Oxford's University Museum – that the new science of fossils can first be said to have become recognisable. It is possible to pinpoint the date rather precisely, too, for on 30 June 1860, while attending the annual meeting of the British Association for the Advancement of Science, a young man called Thomas Henry Huxley had a stand-up fight with a clergyman.

Oxford Encounter. The University Museum, Oxford, England, 51.46N, 01.15W. 30 June 1860

If you walk along Museum Road in the centre of Oxford, you will see an enormous yellow building come gradually into view beyond the lawn on Parks Road. It is magnificent, a tall tower flanked by its pale yellow wings of Cotswold limestone rising nobly into the streets above Oxford. It looks for all the world precisely what it is, a temple to the new theology of the Victorian age: science. It was near here that Thomas Henry Huxley walked on the morning of 30 June 1860. I imagine him pausing and looking across at the same view, his gaze lingering perhaps on the imposing arched entrance with the unfinished adze marks still raw in the stone where, famously, the University had finally run out of money and the Irish labourers had downed tools and gone home. Crossing the road, perhaps he wondered how he had allowed himself to be talked into this, after he had promised himself that he would not become involved in the evolution debate while visiting Oxford. He had made his point to those who would not accept the new world view – particularly Owen – that had come about only the year before when John Murray had published the book that had been so long – twenty years or more – in the writing: Darwin's *Origin of Species*. In later years Thomas Henry Huxley's famous dictum would be that there was no need to reslay the slain.

But the day before, he had seen Chambers and that had been the catalyst for him to remain in Oxford. Perhaps it was because of some residual guilt over his own savage review of the Edinburgh publisher's *Vestiges of Creation* six years before – a book Huxley still regarded as the weakest form of scientific speculation about the nature of evolution – but he had listened rather than passing by as Chambers demanded that he stay on in Oxford and attend the debate the next day. And the next day he found himself standing and staring at the Museum, realising perhaps that his mind had been made up there and then and he had not realised it. His wife Nettie waited for him in Reading, but if he became bored he still had time to catch the four o'clock train.

Huxley was amazed when he entered the Museum. As many as seven hundred – perhaps more – had come to the debate. Expectations were clearly high for this morning's session, even though the previous Thursday when evolution had been discussed the debate had not been spectacular. But he must have realised that the public had come for sport. They had come to see the professors battle it out over Mr Darwin's new-fangled theory. Huxley – Darwin's bulldog as he would come to be branded

– was by temperament a man who did not seek trouble, but neither would he shirk its company if it sought him. When he arrived at the upper gallery he found that the meeting had been moved because the lecture theatre would not seat seven hundred people. It had been switched to the reading room of the soon-to-be-completed library. Inside a corpulent figure resplendent in the purple robes of the clergy sat on a stage: Wilberforce, known to the critics of his smooth tongue as Soapy Sam. Beyond the stage not far from Lady Brewster sat his friend, the botanist Hooker. He joined them and they sat gossiping while they waited for the meeting to start. At length John Draper – the expert in human science from New York – rose and started a lengthy and tedious diatribe on the impact of Darwinism on the intellectual development of Europe. The room grew restless, and soon the students started shouting for Huxley. Henslow, keen to maintain order, moved on and the debate as to the relative merits of the new Darwinian charter for mankind ebbed and flowed but made little progress. Huxley kept his peace, apart from a barbed riposte early on when he admitted with astringent irony that he did indeed hold a brief for science, but had yet to hear it assailed. The crowd bayed for blood. They wanted scientific fisticuffs in this, its new arena. Surely the Bishop would oblige! They shouted for him and eventually, standing and deploying all his orator's skill, Wilberforce put the case for church and creation. His peroration was lengthy and wide-ranging; he criticised the *Origin*'s unphilosophical character and also cited the fact that Egyptian mummies were so similar to modern humans that Darwin's ideas on the mutability of species could not but be wrong. At the end he could not resist a final dig at Huxley, sitting so grave and quiet at the edge of the room. He was reminded said he, of the session only that Thursday in which Professor Huxley had debated the similarities of the brain of man and that of the orang-utan with that other great thinker, Professor Owen. The Bishop now had a question of his own for Professor Huxley: viz, was it through his grandmother's or grandfather's side of the family that he claimed descent from an ape?

Sir Benjamin Brodie the chemist sat on Huxley's other side, and in later years claimed that he had heard the great man utter softly the words 'the Lord hath delivered him into my hands', but no one else heard it. And yet stage-managed to perfection Huxley waited until the noise in the room had reached a crescendo before standing, and waiting calmly and quietly for the noise to subside. He had, said he, listened with interest and attention to the Bishop's discourse, and yet had been unable to discern much or indeed any originality of fact or interpretation within it, except that is for the Bishop's closing remark concerning his own preferences in the matter of ancestry. The question would not have occurred to him, being of no relevance in scientific debate, but it would be churlish not to furnish a reply. And his reply was as follows: if the question was would he rather have a miserable ape for a grandfather or a man highly endowed with influence and mental faculty who then used them for the mere purpose of ridiculing reasoned scientific debate, why then he unhesitatingly affirmed his preference for the ape . . .

The room erupted. The students cheered. This was, after all, what they had come for. Wilberforce paled but said nothing. Lady Brewster fainted, Henslow reddened, Hooker smiled.

Evolution was on the map at last.

The account above is based on the common myth of the debate between Wilberforce and Huxley which is itself based on Frances Darwin's reconstruction of the event in *The Life and Letters of Charles Darwin* published in 1887. Frances Darwin's source was a discussion with Joseph Hooker with some additional material provided by the historian John Richard Green. The truth though is that no eyewitness accounts exist of the debate except for brief comments in two contemporary papers, the *Athenaeum* and *Jackson's Oxford Weekly*. The conflict at the Oxford Museum as popularly portrayed was cobbled together from eyewitness accounts which were over twenty-five years old at the time they were written down. However, Frances Darwin's account was shown to Thomas Henry Huxley, who presumably agreed with it, for it was then reiterated in his own *Life and Letters* (1900) as well as *The Life and Letters of Sir Joseph Hooker* (1918). But in fact no one really remembers what it was that Wilberforce is supposed to have said that so enraged Huxley. And at least one member of the audience – Balfour Stuart, like Huxley a fellow of the Royal Society – is on record as saying that it was the Bishop that bested Huxley on that occasion.

Also, those eyewitness accounts in the *Athenaeum* and *Jackson's Oxford Weekly* make no mention of any furore. Finally, there is the testament of Hooker himself, in a letter he wrote to Darwin only two days after the debate, that it was not Huxley who won the debate but himself and that rather to his own surprise. Hooker says that, although Huxley answered admirably, the rabble in the room was sufficiently noisy that he was not easily heard and that Huxley did not address Wilberforce's points directly.

But this is one occasion when the myth is more important than the actuality, for in 1860 Huxley was not yet the great public speaker that he was to become. It seems that it was this very debate that spurred him to the heights of skilled communication that he was to use to such effect in later years. It was the Oxford meeting that crystallised for him a suspicion that he had held for some time: namely that the public had a right to a scientific education. After the Oxford debate Huxley became a regular and passionate lecturer to the working men of Britain. 'Education for all' became his touchstone and perhaps his greatest legacy was the establishment of Britain's primary school educational system.

The reason why this was a turning point not just for Huxley personally but for the new science of fossils as well is because it was at this moment that the new theory of evolution and palaeontology – the science that would do more than any other to investigate it – became more than intellectual curiosities and started becoming two of the fundamental underpinnings of modern thought and culture. Palaeontology had pre-viously been a debate about abstracts, but now it found itself thrust firmly into the limelight as the science that would finally tell us where our true place in creation was. I like to think that during his visit to Oxford in 1860 Huxley felt the tumblers of eternity click and lock at that moment too.

In later life his essays were collected together and published in a volume called *Lay Sermons and Reviews*. They stand today as some of the finest scientific writing of the nineteenth century. Although all of Huxley's essays are worth reading, there is one that is particularly important for the development of the new science of fossils. It shows clearly just how far Huxley developed his ideas in only eight years after the Oxford meeting of 1860. This seminal lecture was called simply 'On a piece of Chalk' and it was addressed to the working men of Norwich at another meeting of the British Association, in 1868. Much of the south-east coast of Britain is composed of chalk – the White Cliffs of Dover are a good example of its outcrop. In fact, so pervasive is this ancient and compacted ocean ooze across much of southern Britain that it has given its name to the formation as a whole: the Chalk. The Chalk is packed with billions upon billions of microscopic fossils and it was these that Huxley chose as his starting point that day in Norwich when he embarked on an odyssey through the entire geological history of the British Isles. He held his audience enthralled with tales of ancient seas whose sluggish waves had once rolled where now they sat; from the Chalk of the Cretaceous period he took them through more recent ages of the Earth – through the Palaeocene just after the dinosaurs had died, through the Pliocene when elephants roamed where the Norfolk Broads now run – to the present day where the glacially eroded county of Norfolk now sleeps comfortably on England's far eastern bulge. He made it relevant to the present by telling them that their new knowledge of the Cretaceous ocean which deposited the Chalk was informed by the even newer knowledge that they had wrested, only ten years before, from the modern-day Atlantic through the survey of HMS *Cyclops*, the British naval vessel that had performed the survey of the route for the first transatlantic telegraph cable.

The fossil groups that Huxley used to illustrate his thesis are two that we shall come to know very well – you may think *too* well – within the following pages: the planktonic foraminifera and the coccoliths. Both of these groups are as tiny as *Diplodocus* was huge – the largest of the planktonic foraminifera never exceed half a millimetre in diameter and the coccoliths are ten times smaller still. In fact, the former are known technically as microfossils and the latter as nannofossils. This size range from nannofossil to dinosaur – the unimaginably tiny to the incredibly huge – is itself a metaphor for the spectrum between the old and the new science of fossils. They are the border markers within whose boundaries a quiet revolution is taking place. The microfossils and nannofossils are central to one of the newer disciplines of palaeontology: palaeoceanography, and it was with his lecture on the Chalk that Thomas Henry Huxley laid this discipline's foundations.

But the origins of the new science of fossils are not solely rooted in the musings of Darwin's bulldog and his preoccupation with the microscopic fossils that would eventually open up new vistas of analytical palaeontology; the classical arena of palaeontology – dinosaurs and other large vertebrate fossils – was just entering its own renaissance, 6000 miles from Norwich and as a direct result of one of the greatest engineering marvels of all time.

Como station, Wyoming Territory, 41.50N, 106.10W. 19 March 1877

The wind thrashed the rain against the thin window-pane. And when Bill Reed stopped pacing he could feel the station building heave in the gusts from the Nebraska plains that lay far to the east. The wind! In Wyoming the wind was a force to be reckoned with. It hurled itself across the miles of undulating scrub and grass and it never stopped. Scratching and searching with implacable energy for some point of entry where it could get in to chill you to the marrow and remind you that out here it was master. And there was the noise, too, the draft coming through the narrow planks of the ill-fitting door whistled up and down tormented octaves and set his teeth on edge. This was March with the thaw already underway. When the snows came back in October, that same wind would pick up the flakes and drive them in sculpted, thirty-foot high drifts into the lee of the hillocks that littered this tormented terrain. Yet the recently built station house was tolerably warm, for in Wyoming coal was abundant. They were only a dozen miles west of the halt at Carbon where the richest seam in all the 4000 mile length of the Union Pacific Railroad had been found. Outside, the station platform was deserted. William Reed was one of two men at tiny Como station that day and that was just about as busy as it ever got. Como station was just a watering halt for the locomotives after they'd panted their way up from Nebraska into the foothills of the Rocky Mountains. Rough and ready was the watchword at Como in those days.

Looking closely at the platform, Reed could still see the adze marks made eight years earlier when Uncle Pete had pushed on from Laramie towards the Mormons and the West. Uncle Pete, otherwise known as the Union Pacific Railroad, everybody's uncle in this part of the world, maybe more or less beneficent depending on his whim, but the only civilising influence – which is to say method of communication – in the Territory of Wyoming in the year of our Lord 1877. And it had to be said that Uncle Pete didn't pay too well, which was the heart of William Reed's problem. Which was why he was about to augment his income. At the tiny desk beside the window his companion Bill Carlin sat frowning, writing painfully in his uneducated hand the letter that they hoped would turn the trick. Reed would have preferred to write the letter himself, but Carlin had insisted. So now Reed waited. In the corner beyond Carlin, beside his Sharps carbine, was the bone. Huge, heavy – and hopefully the route to folding money. But for Bill Reed there was something more. When he gazed at the bone his imagination soared. The leg bone of a dead giant – *Megatherium* he was sure. From the Tertiary era. Something not now to be found anywhere in the United States or its Territories, something not known anywhere in the civilised world. A relic from the past.

However Carlin, as Reed was beginning to realise, did not share his enthusiasm. His interest in the bone as far as Reed could tell was purely financial. Which was why he had insisted on writing their pseudonymous letter with its gauche conclusion. Reed's lip curled in contempt. Greed. Pure and simple. But he couldn't even really condemn Carlin for that, with the pay they made and the long, long hours of solitude, cash – and the prostitutes back down the line in Cheyenne – were about the only things that kept a man going. Como was just about the smallest, saddest station on the bleakest stretch of godforsaken high iron in the world, ten miles east of Medicine Bow in the borderland of darkest Wyoming.

And there was the other thing: Carlin may not be his type of person but, here in the wilderness with the next train not due till the day after tomorrow, he was stuck, and it wouldn't do to fall out. For Carlin was the station foreman while Reed was merely the section head for the tiny chunk of line between here and Rock River, and it was within Carlin's gift to make him wait outside the station house if he desired. And that meant sitting in the wind.

Reed let Carlin write the letter and conclude it, as they had agreed, with the enigmatic signatures: Harlow and Edwards.

Americans have a fondness for the term 'badlands'. It means any region of the West (and there are many) where one wonders just what the land could possibly be good for. In the case of Wyoming these lands are the desiccated, tortured and eroded remnants of lakes and river systems that in the fifty-odd million years since the original uplift of the Rocky Mountains have been scoured and sculpted until the land is littered with the arthritic remains of buttes and bluffs. This is a land of spectacular and unique scenery. This is the land of *Shane*. Remember the way that the mountains come up out of the plain in the film? That isn't a bad glass painting, that is really the way it is out there. The scenery of *Shane* is a true testament to the scenery of Wyoming, and the scenery of Wyoming is a porthole into the new science of fossils. Travel south on Route 191 towards the metropolis of Farson, Wyoming – one filling station and a telephone kiosk at a crossroads, blink and you'd miss it – and there to the west, across a perfectly flat plain of sagebrush are the Wind River Mountains marching in a rule-straight line to the south-east and coming up sheer out of the alluvium. Nobody forgets Wyoming.

For the palaeontologist the 'cowboy state' holds a special mystique. It was, after all, in Wyoming that part of the new palaeontology was founded, and it is in Wyoming where – 130 years on – the new palaeontology has made some of its most spectacular advances. In the south and west of the state are the Green River badlands, the remnants of a 4000 square mile fossil lake that covered the land 40 million years ago and which is now the home of Fossil Butte National Monument, source of some of the finest fish fossils in the world. In the north and north-east are the Powder River and Big Horn basins; massive sedimentary infills that contain some of the most important fossils of Cenozoic mammals in the world.

But it is the fossils of the south and south-east of the state that put Wyoming firmly on the palaeontological map. For it was here that Carlin and Reed found dinosaurs and unleashed a feud that was to rock the American scientific establishment to its foundations. Stand on the route of the Union Pacific Railroad and you can still feel them stalking the land – like the fossils they found – the brooding presences of nineteenth-century dinosaur hunters Othniel Charles Marsh and Edward Drinker Cope.

What is it about Wyoming that makes it such a Mecca for the palaeontologist? The awesome suddenness with which the Wind River Mountains come up out of the ground is a clue. For Wyoming has one of the most complex geological histories of any place in the world. It is this complexity that has resulted in the diversity of fossils to be found there: fish, advanced mammals, primitive mammals, snails and, of course, dinosaurs.

In the years before the Cenozoic, Wyoming was at approximately the same latitude as it is now. The North American continent has always been big to float, and it has not moved far in the past 200 million years. But one thing was very different. Wyoming, like the states of Kansas, Nebraska, Arizona and New Mexico, was covered by an ocean that stretched from the Yucatan (Mexico) in the south to Hudson Bay in the north, the Western Interior Seaway.

This was the era of the dinosaur, the Mesozoic, composed itself of three subdivisions: the oldest, the Triassic, then Jurassic, then the youngest, the runt of the litter, the Cretaceous. The greenhouse gas, carbon dioxide, like oxygen was higher than today (some say by up to eight times) and temperatures were warmer, particularly at the poles. As a consequence, sea level was an average of 75 *metres* higher than today. The American West was flooded. In the Western Interior Seaway, mososaurs and plesiosaurs swam; on its swampy shores giant herbivorous sauropods crunched their way through primitive cycads and ferns. Lords of creation were the carnivorous dinosaurs, *Tyrannosaurus*, *Allosaurus* and *Velociraptor*. This was an ecosystem that had evolved over 200 million years of Mesozoic history. The entire tenure of the Cenozoic mammals would fit comfortably into it three times. As for man, let's not speak of him yet. Mankind is the gatecrasher at the party of life; the friend of a friend who arrives late, doesn't bring a bottle and is the last to go.

But then the hammer of the universe hit the Yucatan peninsula. In the oceans the plankton died, the ammonites gave up the struggle and up at the summit of the food chain the dolphins and whales of the Mesozoic seas, the mososaurs and plesiosaurs, starving, died too. On land, the prototype of a nuclear winter cast a pall around the globe, the balmy Cretaceous gave way to the arctic grip of the Cretaceous–Tertiary boundary climate. The world died.

In the earliest years of the Cenozoic the remnants of life that were left picked themselves up, dusted themselves down, looked around and started again. But North America was stirring now, and the Rocky Mountains were beginning to emerge from the continent's granite core. The Western Interior Seaway and its two spurs, the Hudson and Labrador Seaways, rapidly drained of ocean. In Wyoming the several sub-ranges that comprise the Rockies – the Medicine Bow, the Sierra Madre, the Owl Creek, the Wind River – were thrust up in the first ten million years of the Cenozoic. And in the south-west corner of the state, the upthrust of the Uinta Mountains and the Wyoming range formed a natural catchment in which a fossil lake, Lake Gosiute, formed. This intermontane (literally 'between mountain') puddle was a giant. Covering 40,000 square miles Lake Gosiute was contiguous with a similarly sized lake in Utah: Lake Uinta.

For the next 50 million years these new mountains were ground down by erosion. The basins between them – the great intermontane basins of the North American Rockies – gradually filled with the detritus of wind and water until the tips of these once-proud mountains poked out of the detrital plain like the roofs of cars abandoned in a flooded quarry. And then in the Miocene came rescue. For reasons unknown these mountains were uplifted once more (in an event known as the exhumation of the Rocky Mountains) and the erosional cycle started all over again. The net result of all this tectonic upheaval was the formation of an alien, battered, but strangely beautiful landscape, a landscape littered with the remains of animals that had lived before and during these events but were denied a quiet grave by all these contortions.

It was the Union Pacific Railroad as it pushed westward in the 1860s on its epochal journey towards transcontinental communication that finally despoiled their uneasy rest. For the bones of dinosaurs and Mesozoic mammals lay scattered about, there for the taking, and in the acquisitive, heady years of late nineteenth-century America two people in particular were ready to do just exactly that. The story of modern vertebrate palaeontology really starts with these two men who controlled access to the vast fossil fields of the Rocky Mountains in the final decades of the nineteenth century. All of which brings us back to Como station in Wyoming Territory in the year of our Lord 1877.

It is difficult now, in the aftermath of a thousand cereal packet promotions and two Spielbergian rollercoasters (*Jurassic Park* and *The Lost World*) to envisage a world not underpinned by the concept of the dinosaurs. However inaccurately (for they were creatures of the Mesozoic, the 'middle age' of the Earth), they are the basement of our world view; this group of animals, easily more diverse than our own family of great apes, ruled the Earth for longer than the entire span of Cenozoic time and then shuffled off this mortal coil at the K–T boundary. But in the mid-nineteenth century they were a scientific curiosity, nothing more, pursued by diffident intellectuals such as the British anatomist Richard Owen, a great adversary of Huxley's. And then the University of Yale founded the prestigious Peabody Museum of Natural History and appointed a young man of thirty-four as its first director. Just how Othniel Charles Marsh landed such an appointment at such a tender age is in itself a curious story. Marsh was the nephew of George Peabody, a man who had made millions out of banking. Young Marsh persuaded Uncle George to fund him through Phillips Academy at Andover, Massachusetts and then through graduate work at Yale. Following the time-honoured tradition of well-heeled young American males he then travelled extensively in Europe where, rather than doing the art galleries and possibly finding himself a wife as was the prevailing custom of the time, he assiduously toured the museums and universities of Europe buying (yes, buying) fossils for the great museum that he dreamed of founding in America. And then when he returned to the States, phase three of the plan was to persuade Uncle George that Yale really needed a state-of-the-art natural history museum. Uncle George, by now presumably mesmerised by his energetically fast-thinking and quicker-talking nephew, obligingly disgorged the necessary cash – at which point young Othniel helpfully reminded his benefactor of the requirement for a dedicated professor to run the Museum. Someone well-versed in the intricacies of zoology and palaeontology obviously, and somebody who knew how the Yale system worked . . .

You can guess the rest, and so it was that in 1866 Othniel Charles Marsh became Yale's first Professor of Palaeontology. This set-up is not as unique as you might suppose, and nor is it some antediluvian artefact from a bygone age of academic patronage. Far from it, it's happening today again as government funding levels nose-dive and universities feel the cold draft of fiscal mortality around their ankles.

It took Marsh a few years to get settled in, what with overseeing the construction work and thinking about staffing his cathedral to the science of fossils. But in 1869, just a year after Huxley had lectured on the Chalk in Norwich, it was time to think once again of vertical mobility in the academic world and so O.C. (as he liked to be known) headed out to Kansas for the annual meeting of the American Association for the Advancement of Science. In the grand tradition of good conferences before and since, the meeting ended with a good thrash; take your choice of three – visit the coalfields of Rock Island, visit the Lake Superior iron mines, or swallow your pride, check no one is looking, accept a little largesse from private industry and take a little trip to see the wonders of the cutting-edge technology of the day: the tracks, locos and railhead of the Union Pacific Railroad Company.

Marsh took the freebie and headed west on the high iron, dining in baronial splendour in the private coach of one General Casement and reaching the railhead at Benton, Wyoming three days later. *En route* he stopped at some godforsaken, blighted wayside halt the name of which he immediately forgot, engrossed as he was with the rather fine claret that Casement kept on board to ease the passage of the West's interminable miles. While there he met the station master who tried to interest him in some bones. But at that time O.C.'s interests lay more in the peculiar shape-shifting properties of the salamanders that had been found in a shallow lake nearby. He collected several specimens to take back to New Haven and then pushed on west. All this was a pity, because the name of the station was Como, the employee who tried to interest him in the fossil fragments was William Carlin, and the fossil was of course a dinosaur bone. It is equally certain that as O.C. ambled his post-prandial way from the station to the lake to collect his salamanders he crossed the fossil fields of Como Bluff and yet never noticed a thing.

The railhead at Benton must have been the inspiration for the phrase 'end of the line'. But here, in this blighted place, Marsh met his destiny. It was rat-shack heaven for the man from New Haven. He might be in the back of beyond amongst the great unwashed, yet he had already seen the potential of the West for science in general and for the science of fossils in particular. He made up his mind there and then. He would focus his professional life here, because here was very clearly where the richest academic rewards were going to be found.

Marsh had seen his opportunity. He had seen the endless prairies and deserts stretching to a limitless horizon and had correctly guessed that this was fertile territory for a man with an interest in palaeontology. He had not yet decided to specialise in vertebrate palaeontology, he was still a generalist. But that decision was a natural outcome of his obsession with the fossil fields of western Wyoming, because they were packed with the fossilised remains of fish, reptiles and mammals. And there was the inaccessibility of it too! O.C. didn't want to spend his life grubbing around in quarries rubbing shoulders with colleagues! O.C. wanted a little slice of territory to call his own. Somewhere he could make his mark, somewhere where everything he did was new, where every find could result in a publication. And the 30,000 square miles of south-western Wyoming territory would suit him just fine. He needed somewhere not attractive to the competition. Wyoming again was ideal; it had disincentives in abundance. Take the weather. To call Wyoming's weather hostile was to say that the British 'lost' at Yorktown. It is not adequate to convey scale. Hostile? In winter the temperatures could plummet to −40°C, in summer they could soar to 45°C. You froze or broiled, take your choice. And in the autumn the wind came, roaring across the sagebrush like an endless herd of buffalo. Your constant companion until spring came and it was time to think about broiling again. In the summer Wyoming heaved with examples of every pest known to science. Midges and flies were constant companions as were the snakes, most especially rattlers. When Marsh came back to Wyoming in 1870, he immediately lost three horses to rattlesnake venom and their [the rattlesnakes'] humming soon became an old tune and the charm of shooting the wretches wore away for all but one, who was collecting their rattles as a necklace for his girlfriend back home.

And then there were the Indians and the bandits. The former were becoming progressively more disenchanted by the way that the US government was relieving them of their ancestral hunting grounds and the latter found easy prey amongst the construction crews and hangers-on attracted by the Union Pacific Railroad. Marsh made it his business to make firm friendships among the army top brass and it stood him in good stead. His expeditions were always accompanied by an armed army escort, the only palaeontologist so favoured amongst all those who would eventually be attracted by the fossil bonanza of western Wyoming. And finally there was the issue of accessibility. There was but one economic route into Wyoming: the Union Pacific. And this was the artery that Marsh decided he would exploit. In those days the deserts and prairies of Wyoming were almost as inaccessible as the deserts of Mars and not therefore (or so he thought) attractive to his competitors.

After he had returned to New Haven, he started planning his big trip West. He had to delay it because of the Indian wars flaring up in the mid-continent but by 1870 he was ready, having solved the biggest problem facing explorers before and since. He needed help, preferably of the inexpensive variety. But inspiration had dawned; he would seek out impoverished students and offer them the chance to make some money. The first Yale expedition of 1870 was led by Marsh and crewed by eight Yale stalwarts. They travelled west on the UP (which by this time had been completed, meeting its counterpart the Central Pacific Railroad at Promontory Point, Utah in 1869), using the line of the railroad as their base of operations. At Antelope Springs, Nebraska, Marsh had already found a fragment of a skeleton. On this trip he found enough new material to describe and name the first pterodactyl, a flying relative of the dinosaurs.

The expedition pushed on into south-west Wyoming. From two of the four basins of the Green River Formation they retrieved fossils of early Cenozoic age. The Green River Formation is the name given to the sediments deposited by the two enormous fossil lakes (Gosiute and Uinta) that existed amongst the peaks of the proto-Rocky Mountains in the early Eocene (55–42 million years ago). The Eocene was a period in the early history of mammals and it is from strata of this age that the initial Yale expeditions of the early 1870s mostly collected. The expedition was a notable success and Marsh organised three others, one each year until 1874 when he returned more or less permanently to New Haven having figured out that he could collect more fossils by pulling the strings of a team of collectors he would establish from Kansas to Utah. Also O.C. began to be concerned that his preeminence in the palaeontological community was becoming compromised, because his ex-friend Edward Drinker Cope was showing increasing signs of interest in the fossils of the West.

Marsh first met Cope in the 1860s. They had even collected fossils together in New Jersey. But after Marsh's almost instant successes in the West, Cope was unable to resist sending his own expeditions along. And therein lay the problem: scientific poaching. Once Cope was sure that Marsh's field area was being successful in producing new fossils he refocused his own research effort on to Wyoming and almost immediately, tensions brewed.

Although Cope did not have the advantage of the universal military escort that Marsh did, he was still successful, naming many vertebrate fossils over the next decade. And it is this business of naming that is at the heart of the matter. For despite the technological marvel that had opened up the fossil fields of Wyoming – the Union Pacific Railroad – in late-nineteenth-century America vertebrate palaeontology was still preoccupied with taxonomy: the naming of new species. The Americans had successfully exploited a technological advance to help them develop palaeontology, but they were behind in their concept of what palaeontology could become. They had no Huxley to visualise the future.

In fact it is hard to overestimate the importance of priority in the naming of species in the nineteenth century. The normal style of nomenclature that palaeontologists and biologists use is called the binomial system. First is the generic (genus) name – for example, *Tyrannosaurus*. The second is the specific (species) name, in this case *rex*. After the specific name comes the name of the worker who named the organism, be it extant or fossil. Naming a new species confers instant immortality on the namer for no matter if later classification systems relocate the species into a new genus, the specific name is immutable and so is its attachment, the name of its discoverer. But species in our world are discrete entities able to breed and therefore reproduce, and this indeed is the technical, *biological* definition of a species: a species is a group of animals that can interbreed and so perpetuate themselves across generations. In the 'fossil record' – the history of life as left in sedimentary rocks – the situation is very different. It is impossible to tell whether a species is 'real' or not, because by being dead, these organisms naturally cannot reproduce. The situation is in practice complicated further as one moves from complex organisms like mammals to simple organisms like the single-celled protists because many of the latter tend to reproduce asexually, e.g. the budding of the *Amoeba*, the binary fission of many bacteria etc. And when the fossils are fragmentary remains of complex land vertebrates and the whole animal is being reconstructed from different parts scattered across the landscape, why then the tendency is to name a new species on the basis of one or some of these fragments.

In nineteenth-century America, partly because of the relative immaturity of conceptual palaeontology there, partly because of the acquisitive nature of the times but mostly because Marsh and Cope had come to despise each other with an abiding fervour the race between them became the race to name new species. It was not enough to name one, it had to be many: Marsh and Cope would settle for nothing less than *multiple* entries in the book of life.

And so from 1871 to 1877 Marsh and Cope fought and bickered, often publicly. It was common for Marsh to accuse Cope of pre-dating his publications in order to claim priority of publication and thus ensure that it was his name that was associated with new species. After 1874 Marsh retired from active fieldwork to Connecticut and hired teams of bone-hunters to do his digging for him. Thomas Mudge was one of the first and most famous of these, later becoming a distinguished palaeontologist himself. Within a year of Marsh hiring Mudge, Cope had tried to poach him. He tried again a year or two later and when that failed he hired his own man, Charles H. Sternberg. From then on the two teams were often found working within a half mile of each other, stumbling across each other in the field and trading insults that often became blows. Wretchedly it was not uncommon for them to destroy fossil material to stop the other gaining it if an existing specimen was already safely to hand. With this type of palaeontological strip-mining in operation it is not surprising that both men's research efforts suffered; Cope, for example, is best known for describing most of the Cenozoic fossil fish of Kansas. Why? Because Samuel Williston, another of Marsh's trusted lieutenants, made sure that all the other vertebrate fossils were removed for the Machiavelli of New Haven!

Then there was the confusion in the literature: by the late 1870s both men had in-house journals. Marsh routinely wrote for the *American Journal of Science*, published by Yale. He had an agreement with the editors that they would priority-publish anything of his, sometimes in the form of an appendix (there is evidence to suggest that they sometimes tired of Marsh's many late-breaking publications interfering with other articles that they had already typeset). This meant that Marsh could uncrate a specimen box rail-freighted in from Wyoming Territory, examine the contents, dash off a quick paper and have it published within a couple of weeks! For Cope there was only one way to counter this, and so in the mid-1870s he purchased the journal *American Naturalist*, and thereafter enjoyed a similarly speedy publication turnaround time. It took years to disentangle their conflicting claims and counter-claims and to unravel the synonymies – the same fossil described twice under different names – that had been generated in their feud. (This task was only completed by the noted vertebrate palaeontologist Henry Fairfield Osborn in 1910.)

All of which brings us back to Carlin and Reed at Como station in Wyoming Territory in the year 1877. This was the year of apotheosis in the war between Marsh and Cope, for in April a retired schoolmaster named Arthur Lakes, a resident of Golden City, Colorado (now home to the famous Coors brewery) had been prospecting for fossil leaves with a friend among the hard Dakota sandstones near the town of Morrison. On one of the hogbacks (a hard, erosion-resistant layer of rock that forms a sharp hillock, particularly common among the foothills of the Rocky Mountains) Lakes came upon a huge vertebra sticking out of the native rock. He wrote to O.C. and received a letter offering to identify the specimen. Lakes replied that he wished to find more specimens before deciding on a course of action, but that they had retrieved another bone, a femur, measuring 14 inches across at the base and indicating an animal of not less than 70–80 feet in length.

Lakes of course was indulging in the time-honoured custom of all hustlers everywhere: by playing coy and yet divulging hints of new specimens he was 'kiting' the price. In April 1877 when he fired off his second missive to New Haven he fully expected that the great man would rush all the way to rural Colorado waving a bunch of greenbacks. And – he got silence. Not a word! It was incredible, the most vociferous half of the greatest scientific feud in America, a man of colossal and acquisitive enmity, a man who routinely referred to his competitor by code (so as not to alert the opposition Marsh had instructed all his henchmen to refer to Cope as 'Jones' in all correspondence so that if any telegraph or letter were intercepted no impecunious railroader or mailman would make the connection and alert Cope) had made – no reply. Lakes was amazed. After all it was no secret that Marsh had paid out many thousands of his uncle's money over the years to those who had located significant fossils for him. And so Lakes decided to do the obvious thing and contact Cope. But he also wondered if the mail to New Haven had gone astray. So the following month he sent a crateful of material, a little taster from bone heaven, to New Haven. The crate arrived in June, by which time O.C. had replied to Lakes, only to be told that some material had already been sent to Cope in Philadelphia. Marsh was aghast; aghast and enraged and terrified that he was to lose the locality to his nemesis. He directed his most trusted lieutenant Thomas Mudge to proceed immediately to Morrison to evaluate the find.

Mudge finally caught up with Lakes at the end of June and soothed the schoolmaster with the universal panacea: cash. All seemed well. The fossils in Philadelphia could be retrieved and redirected to New Haven; after all, Lakes had sent them on approval only. But a hundred miles to the south, in another part of the Rocky Mountain foothills, near the town of Canon City, Act Two was unfolding. Oramel Lucas, superintendent of schools at Fremont County, Colorado and a keen amateur botanist was roaming the hills near his home when he came across some fragmentary bones. He sent them to Cope who identified one as being the largest land animal yet found 'including the one found near Golden City by Professor Lakes'. Oh, revenge must have been so sweet for Cope that day, for by this time he had received the order from Lakes instructing him to forward the Morrison bones to Marsh. He named Lucas' fossil *Camarasaurus supremus*.

Within days of hearing of this find O.C. instructed Mudge to travel immediately to Canon City to examine the material. Mudge reported back that the bones were bigger than those from the Morrison quarries by 10–30 per cent. Bad news indeed for Marsh to whom size so obviously mattered. However Mudge discovered that Lucas too was susceptible to the allure of cash and was already feeling that he had sold the bones to Cope too cheaply. When Marsh heard this he wired Mudge, 'Secure all possible. Jones has violated all agreements.' And so it was that Marsh infiltrated the Canon City operation using superior purchasing power. By the end of the 1870s Cope had spent much of his own vast fortune trying to compete with Marsh. By September of 1877 Mudge's assistant Samuel Williston had been sent to the Garden Park outcrops to assist in the delicate work of extracting the friable bone from the hard, silicified sandstone in which it was embedded. But the Cope team had the best exposures – in clay and marl – and were making better progress so O.C. sent Williston back to Morrison. After a quarry slide that almost killed him, Williston returned to Kansas, narrowly missing the telegram from

Marsh that followed receipt of the letter – the one asking Marsh for cash – from Carlin and Reed.

Eventually Marsh made contact with Williston and dispatched him immediately to Como, where he discovered that the Harlow and Edwards who had signed the letter were really Reed and Carlin. Williston reported back that Como was indeed a find of spectacular importance and Marsh immediately sent instructions that the site should be worked intensively and Arthur Lakes was dispatched from Colorado to assist. For the rest of the 1870s Como Bluff was worked by Marsh's men, often under conditions of great hardship. Reed eventually tired of O.C.'s emasculated interpersonal skills and went into business as a palaeontologist himself. He ended his days as curator of the geological museum at the University of Wyoming in Laramie.

And so the study of dinosaurs, arguably one of the most important public relations successes of the new science of fossils, and a subdiscipline whose development depended on the technological innovation of the railroad, was really initiated by a feud between two nineteenth-century palaeontologists. Marsh and Cope died bitter enemies. At heart the feud was a question of priority in publication and therefore of personal advancement. But the story of Marsh and Cope also serves to date the stage palaeontology had reached in America. At that time, the late nineteenth century, despite the technology that had made the biggest and the best dinosaur fossils in the world available to science, the essence of the subject was still the naming of new species. Huxley's new concept of palaeontology had yet to make much impact across the Atlantic. But if Huxley's ideas represented a turning point in British and European palaeontology, it was because fossils had for a decade already been used functionally in one respect, in stratigraphy: the division of time.

The Hunt for the Ruler of Time

For much of the later nineteenth century as well as a good chunk of the twentieth, palaeontologists recognised three ages of life on Earth: first and oldest was the Palaeozoic, succeeded in turn by the Mesozoic, and then the youngest, the Cenozoic. The phrase 'ages of life' gives a vital clue to the way that the passage of time is traditionally identified in geology, which is by using fossils, or to give it its formal name, biostratigraphy. Stratigraphy refers to the fact that rocks that were once uncompacted sediments (or sedimentary rocks) have a distinct vertical layering with the most recently deposited sediments naturally enough being on top. The prefix 'bio' merely indicates that the different strata are recognised on the basis of the fossils that they contain. There are other types of stratigraphy too, some of which we shall meet in this book: lithostratigraphy means that rock layers are distinguished on the basis of their distinctive colour and textural characteristics; chemostratigraphy means that rock layers are distinguished on the basis of their different chemical characteristics; magnetostratigraphy means that rock layers are distinguished by their magnetic characteristics. And so on.

The fossils used in biostratigraphy are used to measure the passage of *relative* time, and have been ever since the law of superposition (the law that states that the most recently deposited sediments are always the ones on top) was discovered by the Dane, Nicholas Steno, in the seventeenth century. The measurement of *absolute* time is done by exploiting the consequences of radioactive decay, a technique that we shall examine in detail in Chapter 3. Although the law of superposition states that younger sedimentary rocks are always deposited on top of older rocks they may not be found so in the field; folding and faulting often have altered the original sequence so that in extreme cases younger rocks may even lie under older rocks. But they were always *deposited* with the youngest on top.

What is it about fossils that makes them good indicators of time? Fundamentally it is the fact that species evolve. In an idealised case, the characteristic shape (or morphotype as it is known) of, say, a brachiopod species will vary systematically over a period of a few million years as it evolves, yielding a lineage of successive forms each of which can be uniquely associated with different strata in a rock sequence. At its most useful, this sequence of brachiopods can then be found in other places and, even if the local rock type is different, can be used to relate the new sequence back to the original (or reference) sequence. This technique is known as correlation.

In fact, it is not even necessary to use an evolving lineage, as assemblages of unrelated fossil groups can also typify different time horizons. These different time horizons are known as 'zones': a zone can be thought of as the minimum unit of biostratigraphically recognisable time. Of course, in detail, the situation is never this simple; first and most important is the fact that the fossil record is often distorted by periods when no sediment was deposited, thus changing the succession of time-diagnostic fossils. The missing morphotypes indicate the presence of a hiatus.

Recognising time and time-gaps using fossils is the discipline that concerns a particular sub-group of palaeontologists known as biostratigraphers. Biostratigraphers are important for they were the first palaeontologists to actually *use* fossils for something.

Of course there are other limitations to the use of fossils as time indicators. In a perfect world a good zonal fossil would be one that is so widely distributed that it can be found globally. This rarely, if ever, happens.

Living species, as we know, are generally restricted to particular latitudes and continents and this was the case in the geological past, too. Also good zonal fossils should be found in as great a variety of different rock types as possible which means that they are largely independent of the composition of the original sediment (this composition is known technically as a 'facies'). Many fossils are only found in certain facies; brachiopods for example are normally found associated with the limestones that formed on continental shelves in the past while the graptolites (a group that existed at the same time and which we shall meet shortly) preferred deeper waters which deposited sediments that became a type of rock known as black shale. Good zonal fossils should be long-ranging in time with plenty of shape variability so that the biostratigrapher can recognise them easily and yet they should be simple enough so that they are always found whole to maximise the chances of finding often subtle distinguishing features. We can see immediately now how useless the dinosaurs of Marsh and Cope were for this enterprise; dinosaurs are *not* widely distributed, they are *not* found in many different rock types, (in fact they are most commonly found in badlands formed by ancient deposits of streams, rivers and lakes) and their bodies are nearly always found more or less fragmented.

And so it was that when the palaeontological subdiscipline of biostratigraphy was founded in the early years of the nineteenth century the hunt was on for groups that satisfied these necessary criteria. The goal was to identify discrete and non-overlapping segments of time on the basis of the presence (and sometimes absence) of particular fossils or associations of fossils. The goal, in short, was to recognise biostratigraphic zones. Because of the rock-type dependence of many fossil groups it was soon realised that several zonations would need to exist in parallel for the same tract of geological time. For example, the Cretaceous period now has well-established zonations based on ammonites, as well as the single-celled, chalk-shelled planktonic foraminifera and the chalky algal platelets known as coccoliths.

The desirability of biozonation is that it gets around the use of rock itself for dating purposes. Commonly, the colour and texture of rocks change as they grade into each other in the field and this means that rocks themselves cannot normally be accurately correlated across more than local distances.

It was one of the great coincidences of palaeontological science that the three great swathes of geological time that were recognised as the eras of life could – at least to a first approximation – be conveniently zoned using three major groups of fossils. The graptolites are used to subdivide the Palaeozoic, the ammonites the Mesozoic and the planktonic foraminifera the Cenozoic. The story of the pioneers who investigated these groups and subsequently used them to erect temporal signposts in these vast tracts of eternity is one of the most fundamental stories of palaeontology. To tell it, even in the abridged form, it is convenient to follow one of the oldest traditions of geology itself: start at the bottom and work upwards. In this case that means starting with the Lower Palaeozoic, rocks that form the heart of Wales and the Welsh borderland in the United Kingdom as well as much of New England and large areas of China. The graptolite zonation of the Lower Palaeozoic was put together in the nineteenth century and the story of how it was done is nothing less than a Victorian melodrama in the high tradition of Conan Doyle.

Rock rage in the age of Victoria. Trewern Brook, Shropshire, England, 52.44N, 03.05W. 12 January 1898

She could still hear the distant chink of the harness just over the rise behind her, and beyond that the distant rattle of pickaxes as workmen chipped away at the new quarry on Long Mountain. The workings were high enough that she could just see them above the tree-lined lane, where the trap that had brought them from Middletown station still stood. The rust-coloured rock of the fresh scar glinted in the early morning sunlight; to Ethel's eye it was unmistakably Ordovician. A sideways glance at Gerty's face showed that she recognised it, too. How could they not? It was Lapworth's system. Just to the south of them smoke rose in rule-straight lines through the still air from the chimneys of the Buttington brickworks. Ahead, to the east, the frozen field stretched towards the distant wood and the muted rush of running water in its deep valley. Trewern Brook, the geographical border between Wales and England and a potentially complete series within Murchison's Silurian system. A location vital to both women if they were to have their papers read – as they hoped – in Somerset House next year.

At the edge of the field a tree-lined ravine dropped 50 feet sheer to the glimmer of black water among the naked branches. A small opening in the frosty undergrowth was just visible some feet away. Ethel pushed her way past the snagging brambles. Sure enough, a narrow rabbit track led downward at a steep angle.

At the top of the slope, the noise from the brook below was much louder. Looking back, Ethel could just make out Gerty beyond the bracken. It didn't surprise her that she hung back. Dirt and danger were not her style. Suddenly Ethel's riding-boots went out from under her and she sat down hard on the muddy ground with a bone-numbing thump, then slid forward, faster and faster until she was tobogganing at speed. Branches whipped at her as she crashed downwards, then the ground fell away beneath her and she was in mid air. Glancing down, she saw the green tangled mat of vegetation, while in front she caught a sudden glimpse of weak sunlight on black, rushing water. She hit the ground with a whoof and spiralled slowly to a halt in the mud beside the river.

Down in the long tunnel of trees it was almost dark, despite the cold winter sunshine that bathed the ridge above. At first it was hard to see anything, but after a couple of minutes, her eyes began to adjust. The brook here was a narrow gorge, hemmed in on both sides by banks of black rock that glistened wetly. She approached the nearest face carefully, mindful that a stumble would send the icy water over the tops of her boots. When she was within a couple of feet, she stopped abruptly. On the dark surface of the wet-sheened rock of the eastern bank she could just make out a set of tiny dark stripes that looked like brush marks. Pulling out her brass hand-lens and leaning closer she made them out, tiny saw blades clinging to the rock surface. Things that looked very much like one half of a ladder, rungs protruding laterally from one upright.

It was *Cyrtograptus lundgreni*, one of the most important zonal fossils in the Upper Wenlock rock series. Pulling out her geological hammer, Ethel struck the slab a sharp blow on one side, splitting it cleanly in two, exposing a fresh surface to the subaqueous light. It was covered with a confusion of the dark lines. Ethel studied it intently for a few moments, then smiled; a thanatocoenoses – a death assemblage – hundreds of the body fragments known as rhabdosomes covering the slab. She stood back from the rock face and studied it intently again for several moments before moving north. As she walked, she could gradually make out the rock beds in the rock face in front of her which tilted downwards to the left at an angle of perhaps ten degrees. It made perfect sense, of course; if she and Ethel were correct in their understanding of the graptolite zones then *Cyrtograptus lundgreni* was near the top of the Wenlock series here as it was in Scania, hundreds of miles away in the south of Sweden. And the Buttington brickworks, which were perhaps 300 yards away down the brook to their right were known to be lower in the Wenlock. The two facts together led to the conclusion that the geological time represented by these rocks was running forward from right to left and that younger rocks would be found upstream.

For the next hour the two women splashed their way between banks of black shale that grew progressively higher on either side. At intervals they stopped to take samples of the fossils which occurred in groups interspersed by many layers of apparently barren strata. The fossils that they did find were the same mixture that they had found further down the brook towards the brickworks. They were still in the same zone.

By now, the banks were cliffs of vegetation-slicked shale on either side. The boots of the two women were permanently submerged. And then on the right, on the far bank, they spied a thin streak of greyer rock among the trees: a faint seam of carbonate in the darker rock of the shale, so faint that it was almost invisible to the untrained eye, and looking from right to left they saw that it petered out into the same monotonous black of the rock they had already traversed. Gerty's face fell and she turned to go, but Ethel clambered up the slope to a small, disused quarry. Swinging the hammer, she chipped off a small piece of rock, which clove cleanly in two and fell into the scree at Gerty's feet. Glistening in the fragment nearest her was a thin streak of a grey darker than the surrounding rock. Gerty bent closer and felt a thrill of excitement: *Rigidus*, a form that she had seen in Tornquist's collections in Stockholm and had subsequently collected for herself in Scania. She brushed it down and turned her hand-lens on it. Even in the dim light, there was no doubt, *Cyrtograptus rigidus*, previously unrecognised in Britain but common to the sediments of Sweden and the great Silurian sediments of the Baltic. At her cry Ethel ran down the scree slope and the two excitedly examined the find. They clambered up the slope until the weak winter sunlight was stronger through the trees. For another hour they walked again in the thick black sequences of the graptolitic shales, desultorily chipping away at any sliver of rock that looked fresh and *in situ*. This time it was Ethel who shouted, and the two women huddled together in the winter dimness of the Trewern Brook over a fragment of dark rock. When Gerty looked up to meet her friend's gaze she was smiling. '*Rigidus* again!'

The stream section below did indeed encompass the whole of the Wenlock, one quarter of the Silurian system compressed into half a mile of rock.

Fellow travellers on the road of relative time

I first came across Elles and Wood when I was prospecting for carbonate sediments with my colleague Derek Siveter on the banks of the Irfon River at Builth Wells, the self-proclaimed sheep capital of Wales. And, even more recently, we were extending our search for sediments where we could make carbon isotope measurements (*see* Chapter 4) to the very same section at Trewern Brook. I remember well my amazement when it turned out that the most recent reference to the section was still Elles' paper of 1900. The rain had been sifting in a fine drizzle from a leaden sky all day and the going was nothing if not treacherous. As my feet – encased in the latest compound-soled climbing boots with treads that would have looked at home on a bulldozer – went out from under me and I plummeted gracelessly fifty feet down a bramble-covered mudslide towards the vertical drop that would land me in the river, my thoughts turned to those two grand Edwardian ladies and I found myself wondering what it must have been like for them to prospect for graptolites all those years ago. How had they managed in what were probably riding-boots and long skirts? And think of their achievement: together they established the zonation of a significant chunk of Lower Palaeozoic time.

Gerty Elles and Ethel Wood had met when they were under-graduates at Cambridge, reading for Dr Watts' natural sciences tripos. Watts – later to become Professor of Palaeontology there – was a charismatic teacher. He had encouraged their interest in a group of enigmatic fossils that were to be found in certain ancient rock types, predominantly shales and mudstones which were usually devoid of the other types of fossils – brachiopods and trilobites – that were already in general use for correlation. As fossils went, the members of this group, known as graptolites, weren't much to look at; in truth they looked like smears from a 2B pencil. But their importance could not be underestimated, for they had proved to be the decisive weapon in several bitter battles of Victorian palaeontology. It was their role in solving the Southern Uplands controversy as well as the Murchison–Sedgwick feud that stimulated the youthful imaginations of Gerty Elles and Ethel Wood and persuaded them to devote their lives to the graptolites.

Gerty and Ethel both graduated with firsts in the summer of 1895. Gerty went to Sweden to work with several specialists in graptolite classification and came back enthused about the potential of the new science of correlation that was developing. Gerty seems to have been something of a type-A personality, ambitious and determined, with a quick temper and zero patience. Soon after her return to England she landed a lectureship at Cambridge where she stayed for the rest of her life. Ethel's personality and career path though were quite different. By all accounts she was a gentle, sweet-natured girl, without any of Gerty's anxieties or neuroses. And she got the short straw, for while Gerty had a year's tutelage abroad, Ethel went to Birmingham.

The attraction of Birmingham lay in the presence there of the king of the graptolites himself: Charles Lapworth. Lapworth was Professor of Geology and Physiography at Mason College (now the University of

Birmingham). In his lifetime he had been a palaeontological revolutionary, the Che Guevara of the Lower Palaeozoic, his researches settling no fewer than two of the four great palaeontological debates of his age. Born in Faringdon, Berkshire in 1837 Lapworth seems to have been something of a romantic. For example, he was sufficiently a fan of the noted Victorian poet and novelist Sir Walter Scott to move to the Scottish Borders in his youth where he was for several years a schoolteacher and where he solved the Southern Uplands problem.

The Borders of Scotland are made up of great tracts of unfossiliferous rock which in the middle years of the nineteenth century came under the catch-all heading of 'graywacke'. This slaty rock in the Southern Uplands is apparently interspersed by several bands of thick black shale. Black shales have a specific connotation in the sedimentary geological record. They are black because they contain a high proportion of organic carbon, which means that they were deposited under 'reducing' conditions – that is, conditions where there is too little oxygen in the water to oxidise the carbon and return it to the ocean. So the presence of black shales tells palaeontologists that they are dealing with an ocean, or at least a layer of the ocean, that was stagnant. Lapworth arrived in the Borders town of Galashiels in 1866, just a couple of years before Gladstone took his first term of office as prime minister, and spent the next five years wandering around the area trying to understand the geometry of the colossal thickness of the sedimentary rocks there. Previous surveys had concluded that the graywacke in the Southern Uplands was no less than 26,000 feet (more than five miles) thick, but no order to the strata could be discerned because the rock, apart from the black shales, was completely unfossiliferous. So Lapworth concentrated on the black bands. He knew already what the blade-like traces within them were, for graptolites had been described by none other than Linnaeus, the father of biological science, more than a century before. Recognising then an immutable truth of palaeontology – that you have to make the best of what you've got – he decided that the unprepossessing smears of carbon on the laminations within the black bands were all he'd got to go on. He knew from correspondence with his Swedish friend and colleague, palaeontologist Gustav Linnarsson, that progress was being made in Scania using graptolites to put together a zonation of the local Lower Palaeozoic rocks. (The Swedes incidentally have a long and very fine tradition in palaeontological science that continues today.) Influenced by this success, Lapworth made a systematic study of the graptolite faunas in the black shales of the Southern Uplands. The results were disappointing. There was no hint of systematic change in the faunas with vertical progression through the rock strata, a finding that was quite at variance with Linnarsson's results from Sweden as well as the implications of Charles Darwin's recently published *Origin of Species*, the bible on which the science of biostratigraphy depended.

At this point Charles Lapworth should have given up, dispirited. The graptolites, so promising as the fossils that would provide the route into these uncharted deep-water rocks had been found wanting. Their lack of evolutionary and hence biostratigraphic change in the rocks of the Southern Uplands suggested strongly that the graptolites would not, after all, be a universal correlation tool for the Lower Palaeozoic as he had hoped. But to Lapworth their lack of evolution was irksome and inexplicable: why was it that everywhere else the graptolites were known to be a group that showed rapid evolutionary change in the fossil record? What was it that made the Southern Uplands of Scotland different? It could only be that the rocks themselves were so deformed that they were obscuring the record of the fossils.

At this point, Lapworth's thinking had reached a fundamental turning point. On the one hand he could accept that, for inexplicable reasons, the graptolites simply failed to work as stratigraphic indices in this place; on the other, he could dig deeper, choosing to believe that some more complex explanation prevailed and that the graptolites would be found of use. Lapworth chose to dig deeper and spent the next two years mapping the area.

Mapping has a very specific meaning to the palaeontologist. It means noting the position and area of the different rocks that make up a region. This information is later put together as a highly coloured and very detailed geological map. As we saw in Chapter 1, the first geological map was drawn by William Smith and published at the very end of the eighteenth century. By Lapworth's time, about seventy years later, it had become the indispensable conceptual tool for the palaeontologist.

Lapworth walked over every inch of the area with painstaking care, observing the way that rock types appeared or disappeared as he crossed the ground, the way that different rock units thickened and thinned, broke or even folded. And then in 1870 he wrote his groundbreaking paper. He speculated that graptolites did indeed show evidence of organic evolution in this time and place, but that the record of their transformation was obscured because the black shales that contained them had been folded back upon themselves. Over the next few years Lapworth concluded that the black shale bands weren't bands at all, but a single band, and of modest thickness too, only 500 feet at maximum extent. The fact that the graywacke appeared to be 26,000 feet thick at maximum extent was the result of the ferocious tectonic forces that had occurred when Scotland smashed into England, in the time period that we now call the Ordovician. The single 500-foot band had been so folded and contorted that it had been noted over and over again as previous researchers tramped the land. The same thickness of rock had come to be thought of as several separate entities. The entire enterprise of unravelling the geology of the Southern Uplands using palaeontology took Lapworth six years. He published a summary of his researches in 1878 as a paper in the prestigious *Proceedings of the Geological Society of London*, sat back and waited for the fur to fly.

His anticipation of trouble was entirely appropriate. For the Geological Society of London (based at Somerset House where Ethel and Gerty would eventually have their papers read) in those days was the most vigorous forum for scientific debate in Victorian England. And that was really saying something, for Victorian society was on the boil, its love affair with technology and the acquisition of knowledge almost without parallel in history. The closest current analogue perhaps is America. America today is in love with technological titbits – palmtop computers, video-conferencing, nanotechnology, genetic modification – in the same way that Victorians were in love with the technological delicacies of their day – flushing toilets, indoor gas illumination, the kitchen range, the steam engine. The eighteenth century had seen the foundation of several important scientific societies in both Britain and Europe. But a vital difference between the Victorians and ourselves is that in the era of Victoria these scientific societies were frequented regularly by members of the general public as well as notables from the worlds of business, politics and the arts. (It was common for both Palmerston and Gladstone to attend the Geol Soc (as it was, and is, known) to inform themselves on the latest news from the geological front.) As the Oxford meeting which Huxley attended demonstrates well, it was nothing less than *fashionable* to be seen attending scientific meetings! (Contrast this with today's world where science often seems to proceed in semi-secrecy behind the doors of universities and government research institutes. Sometimes it appears as though the entire enterprise of modern science is calculated to exclude the very people who pay for it.)

In London during Lapworth's time there were no fewer than four active scientific societies: The Royal Society, the Astronomical Society, the Linnaean Society (devoted to the natural sciences) and the Geological Society. Since the era of the scientific professional had not yet dawned, these societies were mostly populated by what we would now recognise as enthusiastic amateurs. To a large extent these generally wealthy individuals originally from other professions – banking, the law, the military – formed the core of these societies. In the 1870s, when Lapworth's researches on the Southern Uplands were published, the Geol Soc was dominated by two giants, Adam Sedgwick and Roderick Impey Murchison. Sedgwick was Woodwardian Professor of Geology at Cambridge whereas Murchison was a soldier who had decided to make the transition into science. The discipline that he chose was geology, a pursuit ideally suited to a man with an outdoor bent who hankered after mysteries to solve. Of the two it was Murchison who had money, an independent fortune that had funded his interest in the graywacke rocks of England and Wales since the 1830s. At the time of Lapworth's contribution on the Southern Uplands, Murchison and Sedgwick were among the grand old men of British geology. And it was these grand old men who looked with disapproval upon Lapworth's researches. The younger men – particularly the burgeoning group of professional geologists who were in the process of forming the nascent

British Geological Survey – liked the approach that Lapworth had taken in understanding the structure of the Southern Uplands and agreed with his interpretation of its structure. So why was Lapworth's Southern Uplands work looked upon with scepticism by the old guard? Because for more than thirty years the Geol Soc had been riven by an acrimonious debate and personal feud between Murchison and Sedgwick. Their argument had been going on for so long that nearly everybody had lost track of the details but in the beginning it had been about something very simple indeed: the relative merits of rocks versus fossils for the purposes of correlating strata.

Murchison had a short temper and a big ego. He had started work in palaeontology in the 1820s and within a handful of years, by dint of assiduous fieldwork all over Britain, had made himself an expert. In the process he convinced himself that the correlation of rocks (the discipline of tracing rocks of the same age across distances) was best done by fossils, principally fossils such as trilobites and brachiopods that were commonly found in the carbonate sediments of Lower Palaeozoic rocks. Sedgwick had a different vision. He believed that rocks could be correlated on the basis of their colour and texture characteristics alone (a discipline known as lithostratigraphy). For several years in the 1820s and 1830s the two viewpoints were not in conflict; in fact the two men became friends and in 1839 collaborated in establishing the Devonian system, a geological unit that catalogued the problematic rocks of Devonshire and slotted them into their proper place among the succession of strata in the Lower Palaeozoic.

Earlier in the 1830s, though, both Sedgwick and Murchison had begun working on the apparently uncorrelatable graywacke rocks of Wales and the Welsh borderland. Sedgwick focused principally on the deep-water rocks of North Wales and succeeded in both classifying them and correlating the succession of strata across different outcrops. He named his system the Cambrian after a Palaeolithic tribe that had once inhabited the area. Murchison worked in mid-Wales and the Welsh borderland correlating and classifying in similar fashion but using the abundant fossils found there. He named his system the Silurian, similarly commemorating another ancient tribe of Wales. The two friends published their joint researches on the graywacke in 1835. Neither of them knew that they had lit a fuse that would eventually sunder their friendship.

For the next few years, while Sedgwick and Murchison collaborated in solving the Devonian controversy, other workers continued investigating the Cambrian and Silurian systems in Wales and the Welsh borderland. And this is where the trouble started, for they began to notice anomalies. Rocks which Sedgwick had described as being of Cambrian age started to be found within strata that Murchison had described as belonging to his Silurian system; rocks that had been described by Murchison as belonging to the Lower Silurian were found to be correlatable with rocks that Sedgwick had described as being Cambrian. In short, the boundary between the two systems was found not, after all, to be clear. Neither Sedgwick nor Murchison would give ground – literally – for neither was prepared to lose any portion of the territory where their systems had originally been identified. The battle was bitter, and by 1855 both men were writing elaborate and highly detailed reviews claiming priority of publication as the criterion for retaining this or that outcrop of rock within their own system.

It was into this charged atmosphere that the results of Lapworth's Southern Uplands research dropped like the proverbial brick. It was immediately apparent to all that if the graptolites could solve the Southern Uplands controversy they might also be able to sort out the much bigger problem of separating the Silurian from the Cambrian systems once and for all. And it is to Lapworth that we must doff our figurative caps, for he did indeed rise to the challenge and spent the next several years investigating the overlapping rocks of the Cambrian and Silurian using his beloved graptolites. In the end, his solution to the problem was simple. He erected a third system – and placed it between the Cambrian and the Silurian. In one stroke he had solved the argument and also provided his very own monument in the turbulent history of science. Following Sedgwick and Murchison's lead he too named his system after an extinct tribe of Wales. Lapworth called it the Ordovician.

Lapworth was appointed to the newly established Chair of Geology and Mineralogy at Mason College, Birmingham in 1881, two years after he had proposed the Ordovician system. At his own request the title of the

Chair was soon changed to that of Geology and Physiography (a synonym for geomorphology – the science of landforms). Lapworth used the opportunity provided by his relocation to the Midlands' industrial heartland to concentrate on the nearby Lower Palaeozoic rocks of the Welsh borderland, rocks that are the same age as those that had established his career in the Scottish Borders. At about the same time he started to become keenly interested in the graptolites not just as a group of stratigraphically useful fossils but as organisms in their own right. In 1879, the same year that he produced his groundbreaking paper on the Ordovician system, he also wrote *On the geological distribution of the Rhabdophora*, the paper that investigated for the first time the biology of this group. (Rhabdophora was the biological name for the graptolites in the nineteenth century; today's common term 'graptolite' derives from the more recent formal biological designation of the group, Graptolithina.) Lapworth pursued his biological as well as his stratigraphical studies on the graptolites for a decade and a half, but the dual pressures of scientific productivity and university administration eventually took their toll on his already failing health. When Cambridge produced its two queens of the Lower Palaeozoic world in 1895, Lapworth was more than ready for their help and it was natural that Ethel Wood, with her passion for these fossils, should end up going to work for the king of the graptolites himself.

Elles and Wood's contributions to graptolite palaeontology was the climax to a century that had put biostratigraphy on the palaeontological map. Ethel became Lapworth's devoted assistant in Birmingham while Gerty carved out a niche for herself in Cambridge, eventually becoming the first woman Reader of the University. The two women remained collaborators and very much under Lapworth's influence until the time of the First World War, although Ethel's marriage in 1908 marked the onset of a decline in her active interest in the graptolites. Still, it was during the twenty-year period from 1895 to 1915 – a time that saw the Boer War in South Africa, the death of Victoria, the sinking of the *Titanic* and ultimately the onset of the Great War – that Elles, Wood and Lapworth put together what remains to this day the monumental work on the biology and stratigraphy of the graptolites. *The*

Monograph of British Graptolites is still considered to be the definitive work on the group. Even allowing Elles and Wood to do the groundwork, synthesising Lapworth's data with their own and incorporating the findings of colleagues overseas (particularly in Sweden, for Linnarsson had left his own intellectual legacy), it still took fully those twenty years to publish the monograph. Charles Lapworth died in 1920, soon after it was published, and was buried at Lodge Hill cemetery, Birmingham. In his latter years he had suffered from poor health. Always a man of modest physical strength, the years that he had spent tramping over the Galashiels moors, the highlands of north-west Scotland, and the beautiful Welsh borderland, as well as the stresses endured and engendered by repeated sorties into the bullring of the Geological Society, had simply worn him out.

So what do we now know about the graptolites? For a start it is clear that these fossils, these carbonaceous smears that look at best like lines of glistening ink on a freshly hacked rock surface, are closely related to the ancestors of the chordates, a position that puts them firmly on the branch of life that leads ultimately to ourselves. But after that, the graptolites took a wrong turning and ended up down an evolutionary blind alley. The graptolites belong to the hemichordates – a sister group to the proper chordates and an experiment in evolution that never resulted in any lasting success. Within the hemichordates there are a number of groups that no-one has ever heard of, and still less do people care about. Except that is, for the order Graptolithina. The ancestral graptolites – called dendroids – were sessile organisms, that is they anchored themselves against a hard surface under water – a submerged stone or rock – and grew in much the manner of a modern-day plant. Dendroids are found in the fossil record of the Palaeozoic, but even they are a class of little distinction. The graptolites themselves were the advanced derivatives of the order Graptolithina; they were the M series BMW of the Lower Palaeozoic world. By the end of the Silurian, the age that Murchison fought so hard to place on the geological map, the graptolites had streamlined themselves into the so-called monograptid form; a single stripe with the thecae or vessels in which these colonial animals lived arranged up a single strand – a strand that left but a single stripe on the rock that in ages to come palaeontologists would discover.

Simultaneously with this puzzling simplification of body shape, the graptolites diversified as a group – more and more species being added – until they were, along with certain primitive molluscs, the undisputed masters of the Lower Palaeozoic seas.

That is, until they came to the Wenlock–Ludlow boundary. We now know that this boundary has an age of approximately 420 million years. At this time, for reasons that remain obscure, the graptolites almost became extinct. The number of species reduced by 85 per cent, but a couple of lineages managed to cross the boundary unscathed. So the graptolites survived, yet although their diversity expanded once more they never again reached the dominant status that they had enjoyed in the earlier years of the Lower Palaeozoic: Sedgwick's Cambrian, Lapworth's Ordovician and Murchison's Silurian. The graptolites finally became extinct in the Carboniferous, after enjoying more than a hundred million years as the dominant organisms of Lower Palaeozoic oceans.

It was the fact that graptolites favoured deep ocean waters that made them so useful to our nineteenth-century stratigraphers, for they could be used to zone the graywackes, the otherwise undistinctive rocks that could not be correlated (as Sedgwick never admitted, even on his deathbed) using lithological methods. To this day they are still used, and their zonation has been refined to the extent that an average Lower Palaeozoic graptolite zone is a mere 825,000 years long. Given their distance from us in time – some 250-plus million years – this is less than the blink of an eye to the deep-time palaeontologist. Yet in British sections it is their more spectacular shallow-water cousins (the crinoids, trilobites and brachiopods which were all laid down on the continental shelf of an ancient island called Avalonia as it dipped into a narrow arm of the Iapetus Ocean) that command attention. They still sparkle in the beautiful rocks of the Welsh borderland, and they're certainly worth a visit but, although these shallow-water faunas outshine the graptolites in terms of sheer beauty, they don't come close in terms of usefulness when dividing up time. The graptolites were inhabitants of the open ocean, drifting and evolving on their own as the aeons turned, rarely encountering the teeming hurly-burly of the continental shelves with their rat-race of reef-dwelling invertebrates. And when the great floating colonies died, they sank silent and alone into the deep waters of the open ocean, building up the layers of history that Lapworth, Elles and Wood would exploit. The graptolites are the chronometers of the Lower Palaeozoic world.

Gerty Elles returned to Cambridge and stayed there for the rest of her life. In later years she became profoundly deaf and isolated, rattling her way down the Sedgwick Museum's corridors on her way to an office piled high with papers and rock slabs of graptolitic origin. Dominating the room was her beloved Parkes–Lapworth microscope, an instrument that had been devised by Lapworth himself to facilitate the study of the graptolites. She died in 1960. Apart from her unarguable palaeontological achievements, Gerty was awarded an MBE for services in the First World War where she was commandant of a convalescent hospital for injured soldiers in Cambridge.

Ethel M. R. Wood had died in 1947, after having also collected an MBE for pastoral service, this time in connection with the rehabilitation of ex-servicemen. In both cases it seems, these fine ladies' public spirit was both indomitable and noble. Yet Ethel did not spend her life with the graptolites – the lure of the living was too strong for her bright spirit. She married Dr Shakespeare, a Birmingham don, in 1908 and thereafter gave up her own life in science. While it had been Gerty who had described most of the graptolites they named together, it had fallen to Ethel to draw them for their various papers and for their ultimate, monumental monograph. While I was exploring the Lapworth archive at Birmingham I came across one of her original figures. The quality was stunning, each pen stroke a thing of beauty, the effort each must have required in an age before computers is impressive beyond description. Poignantly, these drawings were eventually rescued from Shakespeare's estate after his death in the mid-1950s.

These days much of my own research is concentrated on the oceans and climates of the Silurian epoch and I spend a lot of time amongst the hills and valleys of Wales and the Welsh borderland. As the rain whistles in from the west in the early autumn, or the sun sets behind Caer Caradoc in a flare of gold at the end of a summer's day, it is hard not to think about these two fine women and the foundations that they laid for the rest of us.

I never fail to do so.

Beer and Middle Time. Ashford Mill, Evenlode Gorge, England, 51.45N, 01.30W. 6 August 1946

The green canopy above was so thick that the sun was almost invisible. Here and there narrow shafts of sunlight speared down and left golden pools on the light-brown crust of the unmade road. Behind him was the lip of the gorge, not that it was much of a gorge really, more like a steep valley that led 60 feet down to the slow-moving Evenlode as it wound its way east and slightly south to where it met the Thames at Iffley Lock. There was no sign of habitation on the steep road, the only manmade object in sight his little Austin parked in the lay-by at the entrance to the quarry.

The man in the tweeds leant back and sighed, the sound unnaturally loud in the silence. The only sound in the last three hours had been the occasional cry of a distant swallow diving the placid surface of the Evenlode. The thought checked him. Dive: the phrase conjured up unwelcome and all-too-recent memories of air-raid sirens, the rumble of bombers and terrifying midnight sorties amongst burning ruination. He shook himself, anxious to shift the memories, but for a moment they clung even in the warm summer air of west Oxfordshire; the torching of London, the despair of the Blitz, the uplifting *esprit de corps* of crowded underground stations at midnight as the streets pounded overhead, but overall the awful, suffocating, dispiriting despair of a war that looked unwinnable from the vantage point of the Ministry of Shipping. Enough in fact to bring a man to the point of nervous immolation – and William Joscelyn Arkell had reached that point and more, wandering further into a wasteland of despair . . .

He fell forward, the moment broken, his face against the shallow, ochre cliff face, sucking greedily on clean air untainted by the smell of smoke. He lay there some minutes, scrabbling with slowing fervour at the hard rock of the Cornbrash, remembering that at last he was safe.

Eventually he pulled himself to his knees again and, after checking with customary diffidence that he was alone, loosened his tie. He had not been cut out for war, in his heart he knew that. Just as he knew that he was secretly grateful for the indifferent health that precipitated the nervous breakdown and had returned him to the relative security of Oxford. But the relief was alloyed with guilt; he had lost too many friends in the war, people that he had spent the last fifteen years working with: Bert Hamblin, the enigmatic T. J. Mason . . . At work, corners echoed with their memories, with their loss. The University Museum seemed a darker place without them. He stood and shrugged off his jacket, then sighed, luxuriating in the sudden comparative coolness. He knew that he'd never be free of their memory.

He savoured the peace for a few more moments, then checked his wristwatch: 11.15. Another hour and he'd drive on down to the mill and then up over the Oolite to Wilcote. From there it was an easy drive down to Ramsden and the Royal Oak. Despite the hardships of continued rationing it was still the best pub in the area. It had been there, a coaching inn on the main Charlbury to Witney road, for nigh on four hundred years; he hoped it still would be around in another four hundred. He shook himself, time to shake the voodoo, but these bouts of depression had become more frequent since the end of the war and the relaxation of fear. Perhaps it really was time to move on. The hints coming from the fens were nothing if not propitious; Arkell realised suddenly that he would not be at all surprised if an offer weren't forthcoming by the end of the year. Enough speculation; it was not his nature. Time for some more of the Cornbrash.

For another hour he hacked at the brown limestone. Every so often he stopped and examined a fresh surface with his hand-lens. The fine grain of the rock was interrupted here and there by tiny white spheres. These were ooliths, nodules of enigmatic origin yet common enough in these rocks to give their name to the two great series that made up the British Jurassic: the Great and Inferior Oolite. The ooliths were inorganic, of that Arkell was sure, yet they seemed always to be associated with the broken shell rubble of the intertidal zone, the edges of the coral atolls. That made perfect sense, for this land had been ocean then, dotted, Arkell suspected, with tiny islands like those of the present-day South Pacific. His stern face softened in the quiet of the deserted quarry as his imagination soared; an infinite azure sea, miles and miles of ocean where once plesiosaurs and other monsters of the deep had lived. Replaced now in their turn by the rolling hills of Oxfordshire. The great cycle of geology. Yet it was not dinosaurs that drew him, nor was it the ooliths, it was the ammonites. With painstaking precision he cleared the friable limestone from around a white protrusion standing out from the cliff face: *Clydoniceras discus*, the zonal form that defined the top of the Bathonian series just as he had suspected. He smiled softly, finally at peace with himself, the last piece of the Evenlode jigsaw, the last piece of the last paper that he would write on the Middle Jurassic period while at Oxford. It was, after all, time for a beer.

Beer was of the essence for William Joscelyn Arkell. He had been born into the Wiltshire brewing family of the same name who to this day produces fine ales which are still on sale at the Royal Oak at Ramsden. Interested in fossils and natural history from an early age, Arkell went up to Oxford in 1922 and took a first in 1925. He stayed on at Oxford, under the professorship of the eccentric W. J. Sollas and started his research into the geology of the Oxford area. It was a propitious move. The Jurassic rock of England stretches in a swathe from Dorset to Yorkshire, and had not at that time been studied in detail. He fell in love with a particular group of fossils. But despite the fact that Mesozoic sediments are younger than those studied by Elles and Wood, the biostratigraphy of this era is more complicated. The reason for this is a lot (but not solely) to do with the geographical distribution of continents. In the Lower Palaeozoic the seas were vast, the continents of our present world grouped together in a south polar supercontinent we now call Gondwanaland. And the graptolites were free to drift across the surface of these vast seas at their whim. But by the time the Jurassic rolled around – 200 million years later (but still 150 million years before the dawn of man) the face of the world had changed. Seas were narrow and numerous and the principal ocean-going community that occupied them had succumbed to a nasty bout of nationalism. This group of animals is the Ammonoidea – an order of molluscs now known to be closely related to our present-day nautiloids. It is their mortal remains – the fossils known as ammonites – that Arkell fell in love with. But it was the nationalistic preferences of the ammonites that had presented great complications for those seeking to zone the Jurassic, for unlike the graptolites, biostratigraphic zonations based on ammonites worked only within a relatively limited geographical area. In fact, the Jurassic world could be divided into two great ammonite provinces in which different groups of ammonites reigned supreme: the Tethyan and Boreal realms, or effectively the equatorial and temperate seas.

Our present-day Mediterranean is the remnant of Tethys, an ocean that once stretched from India in the east to the Straits of Gibraltar in the west. Because the continents of North and South America had not yet collided, the world was effectively ringed by a continuous ocean, and it was in this that the Tethyan ammonoids floated, sometimes near the surface, sometimes deeper in these shallow seas. They were capable of changing their buoyancy characteristics at will. British Jurassic rocks contain remnants of both the Tethyan and Boreal ammonite faunas each of which competed with the other for dominance during the Jurassic. The cause of this faunal give-and-take across a mostly submerged Britain is still unclear but may have been related to changes in sea level. Wales was an island at this time, as were the highlands of Scotland, while in portions of the Jurassic the region around London was a tiny sub-archipelago among the greater vastness of the Tethyan Ocean. In any event, Britain occupied a pivotal place in this battle and it was this fortuitous positioning that allowed Arkell to write his seminal works – the *Jurassic Geology of Great Britain*, published in 1933 when he was only twenty-nine and the *Jurassic Geology of the World*, published in 1956 after he had accepted the offer from Cambridge. Arkell's achievement lies in his correlation of the Boreal and Tethyan ammonite faunas. His zonation is used today, and with refinements added since his death in 1957, the average length of an ammonite zone is now a spectacularly tiny 50,000 years.

A final note: Arkell, in the preface to *Jurassic Geology of Great Britain* voiced a hope that, one day, time would be measured in absolute rather than relative terms. It's a significant comment for in physics laboratories across the world the groundwork to do just this was well underway and the new science of fossils was about to undergo another revolution. Arkell, perhaps the last palaeontologist of the classical school of biostratigraphy founded by Murchison, was about to become an atavism. Before we consider the subject of real-time in palaeontology however, the last revolution in classical biostratigraphy demands our attention, and it took place amid the deserts of north Africa only a decade after Arkell's visit to the Evenlode Gorge. This palaeontological revolution was driven by money.

The Cold War fossil

Today the car has become the West's indispensable tool and toy. We can't imagine a world without the car – although sometimes we might wish it! By the time the fifties came around every teenager in America aspired to their own automobile. Then, as now, America was into conspicuous consumption, the difference being that in those days huge, gas-guzzling engine sizes were unashamedly the norm. To feed America's appetite for gasoline dozens of oil companies had sprung up since the twenties; Phillip's Petroleum, Texas Oil and Gas, American Richfield, to name but a few, and all over the oil-rich lands of Texas and Louisiana fields of nodding donkeys tirelessly sucked black gold from the ground. And by the fifties, too, the hunt was on for new oil territories – particularly in the Caribbean and the Near East. Oil had already been discovered under the deserts of north Africa and management's suspicion was that they had barely scratched the surface – that the reserves there were of truly stupendous dimensions. The rocks where they were drilling had been laid down in a former bay on the southern shore of ancient Tethys in the late Mesozoic and early Cenozoic eras. It was known to the oil geologists as the Sirte Basin.

But the oil barons had a problem. These rocks had no obvious large fossils in them which could be used for correlation – the ammonites having inconveniently died out at the same time as the dinosaurs. And so the search was on for a new fossil that could be used to zone this sequence – and time was of the essence, for every company in the world was trying to exploit these and other reserves (mostly of Cenozoic age) in the Mediterranean region. The first company to crack the zonal code for the Cenozoic would be rich beyond dreams of avarice.

It was in this hothouse atmosphere of feverish greed that a twenty-eight-year-old New Yorker was interviewed for a job when the recruitment executives of Oasis Oil stopped off in Stockholm early in the summer of 1962. He had had some experience of the oil industry in the Texan and Louisianan fields. He was used to roughing it, too: in the bayous and creeks amid the nodding donkeys and the mobile rigs, he'd learned hard living and good camaraderie with the roughnecks. The long hours scooping fresh rock from the wellhead as it was brought to the surface in the drilling mud, the hurried meals of baked catfish cooked in foil atop the steam boilers of the rigs, the icy rush of the day's first bottle of awful Texan beer consumed after the end of a shift . . . His name is Bill Berggren.

Bill was in Stockholm in the early sixties having just spent three years studying for his PhD. In those days, Stockholm, continuing Sweden's long tradition of excellence in all things palaeontological, was pioneering the new science of Cenozoic micropalaeontology: the study of microscopic fossils. The feeling was that the remnants of one particular group – the planktonic foraminifera, inhabitants of the world's oceans since the beginning of the Cretaceous – held special promise for zoning Cenozoic rocks. Bill had worked on the planktonic foraminifera of southern Sweden and Denmark during his PhD studies. He was particularly interested in the effect that the K–T extinctions had had on this group. (We now know that the planktonic foraminifera were one of the marine-dwelling groups that survived the great dying at the end of the Cretaceous, although for a brief interval their diversity was greatly decreased.) He had also, from his vantage point in the periphery of Europe and within spitting distance of the 'Evil Empire' heard stories about what the Soviets had been doing with the planktonic foraminifera. In those days oil exploration in the USSR was the sole preserve of the state-owned oil company, VNIGRI (The All-Union

Petroleum Research Geological Exploration Institute). In the aftermath of the Second World War, the Soviet economy had been in as much of a mess as everybody else's, and the Soviets badly needed the revenue – not to mention the self-sufficiency in the chilly dawn of the Cold War – that came with their own oilfields. So they had set up VNIGRI and in the years since the company had explored all over the republics of the USSR looking for oil. By the middle of the 1950s rumours were emerging that the VNIGRI micropalaeontologists were using the planktonic foraminifera to correlate strata hundreds, if not thousands, of miles apart.

VNIGRI was a remarkable concern in that since the early forties its most able micropalaeontologists had all been women. The triple goads of the awful war on the eastern front, the requirements of Marxist ideology, and Joe Stalin's enthusiasm for labour camps had all conspired to rob the USSR of its manpower. And it was women who were pressed into what had traditionally been men's jobs, including of course oil-field work. And so, when oil was found in the Caucasus and the Crimea, states which we now know as Chechnya, Kazakhstan and Uzbekistan, it was women who were sent to identify the age of the oil-bearing strata and trace its lateral extent. There was much taxonomy of the planktonic foraminifera still to be done in those days and Valentina Morozova, together with her colleagues Nina Subbotina and Vera Alimarina, defined many different species of the planktonic foraminifera and in so doing established the zonation of the Cenozoic rocks of the Caucasian oilfields.

Bill Berggren recognised that something very strange and very exciting was happening in Soviet palaeontology yet he had no idea precisely what. The few papers that were being published were all in Cyrillic and therefore almost completely inaccessible to any American who had gone to college in the fifties. But Bill taught himself Russian and managed a first visit for the purposes of 'literature exchange' in 1958. What he found in Leningrad beggared belief. The fabled VNIGRI head offices were in a bombed-out church with no roof where five hundred palaeontologists – most of them women – were hunched in rows over tiny school desks that stretched back into the gloom. On each desk was a primitive microscope under which rock residues were dissected for foraminifera. And presiding over these serried ranks was the chubby, motherly figure of Nina Subbotina, author of the few papers on the planktonic foraminifera that had made it across the iron curtain.

Relationships were established and in the summer of 1961 Bill managed to join a cultural exchange visit to the USSR for three months in which he met with more Soviet micropalaeontologists both in Moscow and Leningrad. Also, and incredibly for an American citizen in the era of Gary Powers, he joined his Russian colleagues for a summer season's fieldwork in the Crimea and Caucasus. The Russians loved their summer fieldwork among the vineyards and steep hills. It was almost as good as having a dacha on the Black Sea coast. Two things happened that year: first a squad of exiles were landed in Cuba by American military advisors in a fiasco that was later to become known as the Bay of Pigs, secondly Bill heard that Oasis Oil of Tripoli were recruiting in Stockholm. Returning to Sweden, he landed the job with ease. He was just the man they needed to help them with the oil-fields of the Sirte Basin – a basin which had been only a narrow ocean's width from the other sedimentary basins of the Crimea in the early Cenozoic.

In Libya, Bill Berggren and his family lived well. They rented two villas from an expatriate Italian and knocked down the adjoining wall so that the kids could roller-skate between their own territory in the one and their parents' domain in the other. There were no fans or air-conditioning and the 40°C heat took some getting used to after Stockholm. But at the company labs there was air-conditioning in abundance. There was excitement, too. Bill worked on core and cutting samples from a single well – ESSO I1-6 – drilled in the centre of the basin where the sediments were thickest. The literature he had to work with was almost non-existent – a single US Museum of Natural History monograph published by a husband and wife team who had worked in the oil-fields of California (and which mostly concentrated on forams other than planktonics) and a handful of Nina Subbotina's papers. For ten months Bill worked on the Cenozoic of ESSO I1-6, resisting the pressures of a company management that wanted quick results, and when he was done he presented them with a complete zonation and stratigraphy of the Cenozoic of the Sirte Basin. It was now possible to identify rock units from other wells drilled in the area and relate them back to ESSO I1-6. Bill's work was soon vindicated when it was found to apply to other sedimentary successions as far afield as Venezuela and the Caribbean. Others – at first mainly oil-company micropalaeontologists like Walter Blow in South America, Hans Bolli and John Saunders in Trinidad, together with Bill himself, continued the work on Cenozoic planktonics. By this time though Bill was operating from the security of a faculty position at the Woods Hole Oceanographic Institution in Massachusetts where he was to spend the rest of his working life.

Given the Soviet origins of the biostratigraphy of the planktonic foraminifera, it is strange to think that the greatest difficulties over extending the correlation potential of the group came when the western zonal scheme came to be correlated with the fossils of the Caucasus. The problem is as old as stratigraphy itself and is the same one that Marsh and Cope had over their Wyoming vertebrate finds in the last years of the nineteenth century: synonymy.

The parallel development of the planktonic foraminifera in the USSR and the West as time-diagnostic tools had soon led to significant problems of terminology. Having two names for the same microfossil in the West and the East, especially at the height of the Cold War when cross-curtain communication was at its worst, rapidly led to the development of two conflicting zonal schemes that made inter-regional correlation of sediments difficult. The problem persisted right into the nineties. In recent years, Bill Berggren, still active though officially now enjoying a well-earned but rather half-hearted retirement, together with my colleague Richard Norris of the Woods Hole Oceanographic Institution have published the first part of an *Atlas of Palaeogene Planktonic Foraminifera* that has finally resolved the difficulties resulting from the dual birth of planktonic foraminiferal taxonomy and stratigraphy.

So what *are* these obscure organisms that revolutionised Cenozoic palaeontology in the sixties and seventies in the same way that ammonites and graptolites had done for the Mesozoic and Palaeozoic in previous decades? The planktonic foraminifera are the great group of single-celled, chalk-shell secreting organisms that a century ago had been lumped together as *Globigerina*. They are, in fact, the very same organisms that Thomas Henry Huxley started his classic essay 'On a Piece of Chalk' by discussing. They are one of nature's great success stories. To all intents and purposes they are common-or-garden *Amoebae* (as found in all self-respecting garden ponds) which have evolved a chalk shell and decided to live in sea water. The suspicion is that they evolved in the middle or late Jurassic probably in isolated basins on the margins of Tethys. Since then they have undergone four major evolutionary radiations, the original one in the late Jurassic and early Cretaceous, again after the extinctions of the Cenomanian–Turonian boundary, a third time after their near total demise at the Cretaceous–Tertiary boundary, and finally after an enigmatic decline (not obviously associated with an extinction event) that thinned their numbers in the middle of the Cenozoic. The latter two of these radiations are well represented in the fossil record of the deep sea. Of additional value to the student of evolution is the fact that from the earliest days of the study of the planktonic foraminifera, even before they came to be collected in colossal abundance by the activities of the Deep Sea Drilling Project, it was known that they were one of the few groups in the fossil record that exhibited gradational forms between different types. I have included in the plates one of the best known examples of transitional evolution in the planktonic foraminifera, the one that characterised the Palaeocene – the first division of the Cenozoic that marks the time immediately after the death of the dinosaurs – the *Morozovella* lineage.

A word about terminology here: if you think that *Morozovella* is a bit of a jaw-breaker, remember that it is the custom of palaeontologists to honour their colleagues. *Morozovella* of course honours Valentina Morozova, married no fewer than five times in the course of a turbulent life that she devoted to the study of her beloved Caucasian planktonics, she eventually ended up married to the head of the Geology Institute in Moscow. And what of Nina Subbotina? Her name is commemorated by another genus of Palaeogene planktonics, the genus *Subbotina*. Since those early days new techniques – particularly those based on measuring the proportions of the isotopes of oxygen and carbon in foram shells (*see* Chapter 4) have shed light on the lifestyle of these long-extinct groups and it is now known that the subbotinids were a very strange group indeed. Nina would have been vastly amused.

The mid-1970s saw another revolution in planktonic foram bio-stratigraphy: the development of deep ocean drilling through the activities of the Deep Sea Drilling Project. This drilling provided scientists with long sequences of relatively undisturbed microfossils from sediment cores that penetrated deep under the floor of the ocean. And one of the most abundant groups were the planktonic foraminifera.

Today, biostratigraphy still underpins much of the new palaeontology, but it is rarely undertaken for its own sake. Like taxonomy before it, it has become a basic tool required for the more advanced and interpretative aspects of the new science of fossils. For geologists generally (even the many geophysicists who think it fashionable to denigrate palaeontology) the simple truth is that they depend on biostratigraphy to orientate themselves in time, using the same biostratigraphic techniques that two game Edwardian ladies, a fine Oxford don and a gentleman from New York helped found.

The Science of Real-time

Although the science of biostratigraphy was well-established by the turn of the twentieth century, there was still no way to connect the succession of fossil horizons in the geological record to true *numerical* ages (that is absolute dates, calculated in millions of years before present). Real-time still eluded palaeontology. That particular trick had to wait for the establishment of geochronology (as the discipline of determining the numerical ages of rocks and fossils is called) and that, in its turn, was dependent on the discovery of radioactivity. It is no surprise therefore to discover that the father of modern geochronology is considered to be Ernest Rutherford, the young New Zealander who, while working with Oxford graduate Frederick Soddy at McGill University, Montreal in 1902, proposed the theory of radioactive decay. Radioactive decay is the process by which radioactive elements can transform into other elements by the loss of energy in the form of particles or rays. The news that matter could transmute in this fashion caused a sensation in the press at the time, for it seemed to prove that the ultimate goal of the alchemists was indeed attainable. But the rules of this new radioactive alchemy did not allow lead to be turned into gold although, as we shall see, the element lead is central to our story.

Before we can go on we need to arm ourselves with the right

vocabulary to understand how physical and chemical techniques have interacted with palaeontology. This interaction has not only allowed palaeontologists access to the absolute ages of fossils and their enclosing (or time-equivalent) rocks, but also to the way that oceans and climates have operated over geological timescales (*see* Chapter 4). We need to remind ourselves of some of the chemistry and physics that we were taught at school. But for those of you whose school physics and chemistry seems a long time ago, I have added below a primer to help you, which should give you what you need to know to understand the architects of real-time.

An atomic primer

The universe is made up of different types of atoms known as elements. An element cannot be broken down into simpler substances by chemical processes, it is the fundamental functional unit of nature. Elements are atoms and atoms are composed of a nucleus and a shell of electrons which orbit the nucleus. The nucleus is itself made up of two types of sub-atomic particle: the proton which has a positive electrical charge, and the neutron which does not have an electrical charge at all. The orbiting electrons have a negative electrical charge which balances the positive charge of the protons in the nucleus. The number of electrons always equals the number of protons in the nucleus so that the atom is properly 'balanced' with respect to its electrical field. If the electrical balance of an atom is disturbed for any reason then it will seek to regain stability by the gain or loss of one of its component particles.

The elements can be arranged into columns in a table where each column (called a 'group') represents elements with similar chemical properties, that is, they react similarly with other elements. When elements are grouped in this way the pattern that emerges in the rows is one of increasing atomic number (from the lowest in the top left to the highest in the bottom right).

The atomic number is simply the number of protons which are in the nucleus. The number of positively charged protons is always balanced by the same number of negatively charged electrons, and it is the number of electrons (particularly those orbiting furthest from the nucleus) that determines the chemical characteristics of an element. Because they have no electrical charge (hence the name), neutrons do not contribute to the chemical reactivity of an element. But they do contribute to the weight, or more correctly, the mass of an atom. The combined weights of protons and neutrons define a value called the atomic *weight* of an element. Although the chemical properties of elements are controlled by the number of electrons which is a reflection of the atomic number, the number of neutrons in the nucleus can and does vary even in different atoms of the same element. These different types of elements are called 'isotopes'. Isotopes have identical chemical characteristics but they have different atomic weights. It is this difference in atomic weights that imparts useful properties that the palaeontologist and geologist can exploit. There are two groups of elements that can be separated by the characteristics of their isotopes: the stable isotopes do not transmute into other isotopes or elements spontaneously; the unstable or radioactive isotopes can do both by the emission or acquisition of energy in the form of particles (or rays). The term nuclide is often applied to different isotopes formed by radioactive decay.

It was Frederick Soddy who coined the term isotopes in the years following his groundbreaking work on elemental transformation by radioactive decay. In the early days of radioactivity, experiments revealed that ores of uranium and thorium contained various previously unknown radioactive substances. At first these were thought to be different elements and were given unique names accordingly. But it was soon found that if these new 'elements' were mixed with their parent compounds (uranium and thorium) then they could not again be separated by chemical means. As the criterion of chemical uniqueness defines an element, these different substances could not therefore be different elements. Hence Soddy coined the term isotope (which means 'the same place' in Greek) because, by being chemically identical, they occupy the same cell on the periodic table. We now know that radioactive decay can form both different elements or isotopes of the same element depending upon whether the loss or gain of an alpha or beta particle alters the number of protons (in which case a different element is produced) or the number of neutrons (in which case a different isotope is produced).

In 1902, Rutherford and Soddy were interested in the fact that *elements* could transform into other *elements* (this was before it was realised that often they would decay to isotopes of the same element as well). Indeed it was Rutherford and Soddy who formalised the rule that if a radioactive element such as uranium or thorium emits energy then it will change (or transmute) into a different element with different chemical properties and if this 'daughter element' thus produced is itself radioactive then it will transmute again into another element and so on (through a 'decay series') until a stable (i.e. non-radioactive) daughter product is arrived at. The stable end product thus produced is known as radio*genic*, because it

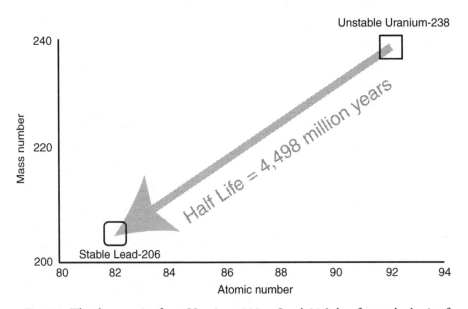

Fig. 3.1. The decay series from Uranium-238 to Lead-206 that forms the basis of time measurement in geology and palaeontology. The half life of Uranium-238 (that is the time required for half the original quantity of Uranium-238 to convert into Lead-206) is 4,498 million years. Several intermediate isotopes and elements are formed and decay away themselves in the process.

has its *genesis* in a radioactive transformation: it is the product of a radioactive decay chain. The radioactive element that Rutherford and Soddy used as the basis of their experiment was thorium, which is closely related to uranium. Rutherford and Soddy suggested that the light element helium was likely to be the ultimate end product of the decay series of both uranium and thorium and only a year later Soddy confirmed that this was indeed so.

The alchemy of ages

Rutherford immediately made the connection between the spontaneous decay of radioactive materials and the potential for measuring real-time in the fossil record. He realised that the rate of accumulation of stable daughter products in minerals that had once borne radioactive materials was, in theory, a perfect natural chronometer. In this particular case, he reasoned, the answer was simply to measure the accumulation of the element helium – the very same element which in gaseous form we are familiar with in children's party balloons – in uranium-bearing minerals. And so, from so simple a beginning was the science of real-time born. Rutherford presented his ideas in an address to the International Congress of Arts and Sciences in St Louis, Missouri in September 1904, just four years after Elles and Wood had presented their papers on the Welsh borderland. So great was the impact of Rutherford's talk that he was invited to give the prestigious Silliman lectures at Yale University the next year. These lectures were published by Yale as *Radioactive Transformations* and contained the first explicit account of the techniques required to calculate the age of a uranium-bearing mineral, based on a *chemical* analysis of the amount of radiogenic helium found in a mineral.

Helium is one of the lightest of the elements and the suspicion existed that some of it might slip away unnoticed from its enclosing mineral prison. In which case, the resulting measurements would underestimate the true age of the rock. Fortunately, helium is not the only daughter product of the decay of uranium and thorium: there is also lead. Bertrand Boltwood, Yale savant and contemporary of Rutherford's, had speculated that this might be a better way of measuring the age of uranium- and thorium-bearing rocks. A chemical method for measuring the lead content of such rocks was generously suggested to Boltwood by Rutherford and the first 'radiometric' lead ages were subsequently published by Boltwood in 1907.

Rutherford was passionate about technological science; in later years a sign would be erected in the Cavendish Laboratory at Cambridge when Rutherford took over as Professor of Physics there which would light up whenever the great man walked in. It read 'Quiet Please!' and referred to Rutherford's tendency to upset the experimental apparatus with his booming voice in moments of high scientific enthusiasm. Rutherford's passion for geochronology inspired R. J. Strutt of Imperial College, London to become involved in the study of real-time. Strutt (later to become Lord Rayleigh) began a comprehensive study of the helium contents of a wide variety of geological materials. The results confirmed the earlier suspicions that helium was not a reliable indicator of the age of rocks. It escaped too easily from its enclosing mineral. By 1910, the same year that Soddy formalised the concept of isotopes, the helium method was largely discredited and attention turned firmly towards lead for making age determinations.

On 6 April 1911, Professor Strutt read a paper before the Royal Society of London. It had been written by one of his students, a man who would become the greatest geochronologist of his age and who would, during a career lasting fifty years, be the one to devise the absolute geological timescale that is still, with only minor modifications, in use today. But in 1911 Arthur Holmes was still only a graduate student and, like so many others, doing much of the hard work for his famous advisor. Yet Holmes' first paper is almost frighteningly prescient; in it he urges the establishment of a numerical timescale in order to date sedimentary successions containing fossils. All of these earliest absolute age determinations were made using chemical techniques. They were crude, and yet they were enough to show that Lord Kelvin's estimate of the age of the Earth (based on his estimates of its rate of cooling published in 1899) was at least a hundred times too low. These earliest measurements were also enough to show that the Precambrian (the time period below the Palaeozoic which is more commonly referred to as the Proterozoic and Archaean) was not a brief interval encapsulating the Earth's convulsive beginnings but in fact represented *more* time than the subsequent Phanerozoic.

But most of the geological community remained sceptical of the promise of geochronology. The American geologist G. F. Becker used Boltwood's uranium–lead methods to date several radioactive minerals from Texas and obtained conflicting results – one even yielded an age of 10 *billion* years, more than twice our current estimate of the age of the Earth and completely unbelievable to a scientific community that only ten years before had thought the Earth to be, at maximum, only 40 million years old. Boltwood cautioned that the rocks that Becker had measured were in an advanced state of alteration, foreshadowing a debate (which I call 'the big D') that was to hang for decades over the whole question of the use of the stable isotopes in the new science of fossils (*see* Chapter 4). But Becker insisted that the whole physical basis of radioactivity-based geochronology was oversimplified, the Rutherford–Soddy law only applied in limited circumstances, and the whole enterprise should be abandoned. By the dawn of the First World War, the notion of real-time in geology and palaeontology was in disrepute.

One of the factors which troubled geologists was whether the rate of decay of radioactive materials (known formally as the 'decay constant') was truly invariable as Rutherford and Soddy insisted. In 1911, physicists Johannes Geiger and John Nuttall showed that the distance travelled by an alpha particle (the nuclear particle with lowest energy and which is therefore the most sluggish) was strictly proportional to the decay constant. At about the same time, the geologist John Joly was working at Trinity College, Dublin on pleochroic haloes, tiny circular scars found in certain minerals around incorporated fragments of radioactive minerals. Joly used the Geiger–Nuttall relationship to show that the radius of pleochroic haloes was always the same, regardless of the age of the mineral. The conclusion was inescapable: the decay constant of radioactive materials was, as Rutherford and Soddy had said, truly invariant. The chronometer of deep time had been discovered.

However, the limitations of the chemical method for estimating geological real-time were becoming apparent. Common rock and its minerals contained so little uranium and thorium that it was almost impossible to measure their radiogenic daughter products using the available chemical techniques. Age estimation was restricted to rocks with unusually high levels of these elements. This prompted a re-evaluation of the helium method, the reasoning being that if too much helium was produced in rocks with a high starting content of uranium or thorium then ordinary rocks with much lower concentrations of uranium and lead would produce less helium which would have lower probability of escaping from the crystal lattice of the rocks. The renaissance of the helium method was pioneered in Germany just before the rise of the Third Reich.

This effort was spearheaded by Fritz Paneth and co-workers in Berlin who successfully detected tiny quantities of helium in iron meteorites by exploiting recent advances in gas-handling and vacuum techniques. The stage was set for a comparison of the accuracy of the helium- and lead-dating techniques. Arthur Holmes and his assistant V. S. Dubey applied both to rock samples taken from two volcanic regions in the north of England. In a state of high excitement they reported their findings in *Nature* in 1929. The results from the two methods were in excellent agreement with each other and furthermore agreed well with the *relative* time information available from the fossil-bearing rocks which bracketed the volcanic layers. Holmes' dream of establishing a universal absolute timescale for the geological record seemed to be within the grasp of science. William Urry capitalised on the success of Holmes and Dubey and in 1936 published the results of thirty-nine helium measurements of basalts (a type of igneous rock) carefully chosen from sites in America, Canada and Europe to encompass the entire Phanerozoic. Urry's timescale was found to agree well with existing uranium–lead measurements. An exultant Holmes incorporated these latest measurements into a revised version of his classic book *The Age of the Earth* and proclaimed 'This test of internal consistency must be regarded as final proof that the ages calculated from lead and helium ratios are at least of the right order and that no serious error is anywhere involved.'

He spoke too soon: disaster lurked just around the corner. Robley Evans was a young scientist who had just arrived at MIT (where Urry was working) from the California Institute of Technology. He had developed an alternative helium measurement method which became known as the alpha–helium method. The heart of the new instrument was a graphite induction furnace which fused the sample into an amorphous blob of molten rock at a temperature of 2000°C and in so doing liberated the total volume of accumulated helium which could then be measured to a very high degree of precision. Evans' new technique was able to show that the electrometer (a standard piece of laboratory kit available in all pre-Second World War physics laboratories) that Urry had used was miscalibrated! The helium dates that Urry had published were all too low, and their agreement with the pre-existing lead-based dates was therefore purely coincidental. To salvage something of his pride, Urry was made a co-author on the paper that was eventually published in *Physical Reviews* in 1939. Two years later it was Evans again, in conjunction with another of the 1939 paper's authors, C. Goodman, who published a review of the radioactivity-based techniques for dating rocks and drew attention to the new methods based on rubidium and potassium, which eventually superseded helium age determination as the technique of choice on common rocks and minerals. The helium method was the last bastion of age determinations based broadly on chemical techniques. Its demise in the early years of the Second World War left the field clear for radiometric age determinations based on a different technique that had its roots firmly in the nineteenth century.

The dark space age of physics

There was a time before nuclear physics really took off which I think of as the dark ages of physics. These dark ages were the last years of the nineteenth century before Rutherford and Soddy changed the world for ever. In the dark ages of physics there was one device that exercised the imaginations of some of the greatest thinkers of the nineteenth century: the vacuum tube.

As long ago as the 1740s, Benjamin Franklin had been experimenting with electrical discharges in rarefied gases in sealed tubes. His friend William Watts remarked, 'It was a most delightful spectacle, when the room was darkened, to see the electricity in its passage.' A vacuum tube is simply a glass tube with a metal plate at each end, one (the cathode) being connected to the negative terminal of an electricity supply and the other (the anode) being connected to the positive terminal. Vacuum tubes continued to exert their fascination in the days of Faraday in the first half of the nineteenth century. He and other scientists of the day were fascinated by the transformation of the electrical discharge as the pressure was lowered and the sparking arc between anode and cathode fattened into a writhing, glowing purple snake; this is plasma, the very stuff the Auroras Borealis and Australis that decorate the night sky over the Earth's poles are made of. Faraday noted that at very low pressures one end of the snake detached from the cathode and began to contract toward the anode. The eerily dull region that it left behind he named 'the Faraday dark space'. These early pioneers also noticed that regardless of the position of the anode the glass began to phosphoresce directly opposite the cathode.

A major advance occurred in about 1857 when the German Heinrich Geissler developed two radical innovations: the first was to insert the electrodes as part of the tube-blowing process; the second, perhaps even more revolutionary, was the development of the mercury diffusion pump which allowed lower vacuums than ever before to be imposed on the space inside the tube. By 1869 another German, Johann Wilhelm Hittorf, had shown that, in a very high vacuum, Faraday dark space expanded until it filled the whole tube and he also confirmed that wherever the anode was placed relative to the cathode the phosphorescence of the glass would still occur at a spot directly opposite the cathode. These rays were subsequently named 'cathode rays' by the physicist Eugen Goldstein in 1876.

In 1879 the Englishman William Crookes showed that by inserting a specifically shaped piece of metal in the tube (he used a piece cast as a Maltese cross) a similarly shaped shadow was cast on the tube's end wall; cathode rays could be blocked. Crookes also deflected cathode rays using a magnet and contended that they were not, as others had suggested, light waves, but negatively charged particles which had a mass. In 1886, Goldstein demonstrated that positively charged particles could be detected in the region behind the cathode travelling in the opposite direction to the cathode rays themselves, if a hole had been cut in the cathode. He named these 'positive rays'. It was J. J. Thomson, Cavendish Professor of Physics at the University of Cambridge and a man of underwhelming technical prowess who set himself the task of unravelling this rat's nest of particles, rays, cause and effect. To separate these effects it was necessary to disentangle their energy, mass and velocity. He used Crookes' tubes at very high vacuums and with very high discharge voltages. By means of a special arrangement of magnetic and electrical fields he was able to deflect the beam (observed as a phosphorescing spot on the end of the tube) and by measuring the amount of charge needed to deflect the beam by a given amount was able to calculate the beam's velocity and charge. He concluded that cathode rays were composed of particles smaller than an atom. It was in fact the first inkling that particles smaller than an atom might exist. Thomson also found that the cathode ray particles were independent of the material of the cathode: cathode rays were one of the fundamental particles of the natural world.

The nineteenth century saw the spectacular breakthrough in physics which was to be so comprehensively exploited by Rutherford: Henri Becquerel's discovery of radioactivity in 1896. It soon became clear that there were different forms of several of the radioactive elements. These were the different forms that were named 'isotopes' by Frederick Soddy in 1910 because they had the same chemical properties and therefore occupied the same slot in the periodic table. It was thought that the phenomenon of isotopy was indivisibly related to the phenomenon of radioactivity.

In 1897 Wilhelm Wien had succeeded in deflecting positive rays with electrical and magnetic fields just as Thomson had done with negative or cathode rays. Wien concluded that the positive rays were atoms or molecules of gas within the tube that had acquired a positive charge when an electron was knocked out of them by a passing cathode ray on its way to the anode. The positively charged particles were then attracted to the negatively charged cathode and if not stopped (for example if a hole had been made in the cathode) would continue beyond the cathode.

Thomson, too, was curious about the nature of these positive rays. The problem was that positive rays were very hard to study because after passing through the hole in the cathode it was not long before the positive ray particle would hit another gas molecule and gain or lose another electron which would render it electrically inert. Thomson once again exploited the advances in vacuum technology. By imposing really high vacuums in the tube *behind* the cathode, he ensured that there were no other atoms for the positive rays to hit as they entered that region.

So it was that in 1913 Thomson showed that the positive rays formed by the ionisation of neon gas exhibited two different mass/charge ratios, one 20 times and the other 22 times that of hydrogen atoms. There were no other detectable mass/charge beams in the tube and so Thomson concluded that there were two isotopes of neon. The era that Rutherford would later famously describe as the 'great age of projectile physics' was born. It was also, as we have seen, the dawn of the great age of the application of isotopes to geology and palaeontology, yet it was not Thomson who went on to exploit the positive ray apparatus, but his almost unknown young colleague who, within ten years of the end of the First World War would win a Nobel prize for his work on isotopes.

The Machine that Bent Atoms, Cambridge, 52.12N, 00.07E. 23 December 1919

Cambridge's wind has a special quality at any time of year, and in the winter of 1919 it chilled Francis Aston to the marrow as he cycled swiftly along King's Parade towards the Cavendish. The fat cobbles of the road were icy and his already erratic front wheel threatened to slip sideways at any moment and deposit him unceremoniously on his flannelled backside. But he continued rattling his way past King's with its brightly painted Victorian post-box and imposing arched lodge. The College's chapel glistened in the cold sunlight, proclaiming the College's vaunted musical aspirations – as well as its wealth – to the world. At the site of the sheer, cold sides Francis' thoughts turned fleetingly back to his own comfortable college rooms further back up Trinity Street at St John's. The fire in the saloon would be bright now that his scout had been in to turn it, but within minutes she would find that for the third time that week he had forsaken his first cup of tea and the muffin on which he customarily breakfasted.

But it was worth the early starts, for the apparatus was almost completed. Down narrow Free School Lane he cycled, past the bay front window of his laboratory, and then turned hard left under the peaked archway of the Cavendish itself. Standing his bike against the wall of the tiny courtyard within, he then walked back into a shadowed doorway and past the entrance to the steep steps that led to the famous tower room where Rutherford was revolutionising science. He walked down the narrow passage and into his cluttered office. The stone floor was covered in a fine patina of glass dust that crunched under his brogues, and the wooden benches that surrounded the room were covered with a hazardous confection of old glass tubing, burnt-out electrical valves and fragments of bent wire and twisted metal. In the corner of the room a glass-blower's torch still sputtered yellow with the rich coal gas that fed it. From its brass valve another length of flexible metal hosing led to the bulbous oxygen tank. With the two gases combined, the flame was hot enough to melt the hardest silicon glass. Aston rubbed the tips of his fingers against each other in rueful memory, they were baked hard by more than one bruising encounter with white hot glass. But, like the early starts, that too was worth it, for on the bench in the middle of the room, four foot high at the maximum extent of its curvature and dominating that narrow space stood the machine.

Aston circled it carefully, admiring for the thousandth time the smooth sweep of the glass flight tube, the uncompromising gunmetal grey of the frame that supported the pumps and the sprawl of frayed wire that covered the entire system like so much confetti. He had had to think of a name for this contraption now that it was almost ready and only the previous night had decided on 'mass spectrograph'. What could be simpler? This was an instrument that ignored the complicated chemistry of atoms and reached straight for the heart of the matter: their weight. But in the arcane world of physics Aston was only too well aware that weight was too imprecise a term; the particles that his machine separated were differentiated by mass. Drawing a hand along the machine's harsh metal surface, he reflected on the analogy that he had used only last night in the senior common room to explain the machine's operating principles. It was based on something that he had had seen only a week before at the college's sports-field out on the Madingly Road. Two runners – undergraduates – had been panting hard as they rounded the final curve towards the finishing line. It was an uneven match, as one must have been a good head taller than the other and weighed at least two stone more. He was struggling, panting far behind his smaller companion who, head down, was running like a thing possessed. As the lighter man had flown round the final corner he had eased gently and unknowingly towards the outer marker of his lane so that as he entered the final straight he was running at the outer edge of his lane. Not so the other runner. Aston had seen that as the heavier runner entered the corner, puffing and winded but still gamely making good time, he had not deviated nearly as far as the lighter man; his heavy gait had kept him centred firmly almost in the middle of his lane all the way around the corner and down into the final straight.

And last night, as he stared into the flickering flames in the hearth and searched for the words to explain his invention to a classicist and a historian, that image had come back to him. The masses he was measuring, he had said, were like the two runners; because they had different weights, they were unable to stay the course with equal precision. The smaller runner – he of lighter mass – had drifted off to the edge of his lane – deflected by the combination of his own velocity and the need to turn the corner; the heavier mass had not deviated as far. And Aston's apparatus had a bend in it, like the sportsfield at Madingly Road, and, just like the runners, the atoms or molecules in the machine's flight tube would need to turn the corner. However, in his machine, the corner would be negotiated under the deflecting influence of the magnetic field that awaited the particles just as they eased into the turn. And when they left the corner they, too, would occupy different positions within their own lanes. They would have been separated by the different responses of their mass.

After this pronouncement Aston had felt his audience stare at him, felt their amusement. But he had not cared, for a machine that could separate big atoms from small atoms, just by bending them through a magnetic field, would change the face of science. Of that he was sure.

The shadow of eternity

Aston's invention was a development of Thomson's positive ray apparatus. The mass spectrometer (as it is now known) stands head and shoulders above all other technological innovations in its impact on the science of palaeontology. It is ironic though that when it was first constructed it was seen as a vital tool for investigating the then burgeoning science of nuclear physics, its applications in geology and palaeontology undreamed of. Yet without the mass spectrograph and its modern-day descendants the science

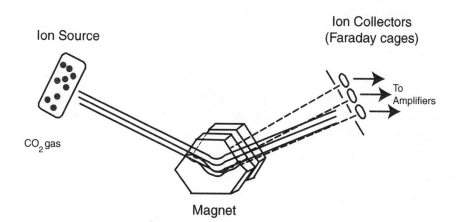

Fig. 3.2. Principle of the mass spectrometer. The three beams of carbon dioxide formed by ionisation of the gas in the machine's 'source' are separated as they pass between the poles of a magnet. The diverging beams are then collected in Faraday cages and the number of 'hits' amplified electronically.

of geochronology would not have been able to advance, and the modern subdisciplines of palaeoceanography and palaeoclimatology could not have been born. Today the mass spectrometer is found in all serious palaeontology laboratories across the world, and it and its variants have made possible the ability to date fossils, as well as to use them to reconstruct the environment in which they lived.

Francis William Aston was born on 1 September 1877 in Harborne, now a suburb of Birmingham. His father was a metal merchant and his grandfather the director of his own engineering company. The atmosphere that Aston grew up in celebrated the mechanical marvels of the Victorian age. At the age of fourteen his father sent him away to Malvern College and then, in 1894 he joined Mason College, Birmingham (where Charles Lapworth had already been Professor of Geology for over a decade) where he studied chemistry and physics. He left Birmingham for a time to work as an industrial chemist for a brewing company but returned in 1903 with a scholarship to study the properties of discharge tubes. Soon he had discovered another dark region within the discharge tube very close to the cathode. He named this area the 'Aston dark space'. In 1909 he left Birmingham to join Sir J. J. Thomson at the Cavendish Laboratory in Cambridge and assist him in studying positive rays. In terms of technical expertise, Aston was everything that Thomson was not and so together they made an excellent partnership. They calculated the mass of various atomic particles by firing them through crossed electrical and magnetic fields and tracing their parabolas in primitive cloud chambers. Noticing that neon described two parabolas, Thomson instructed Aston to try to isolate them. By 1913 Aston had partially succeeded by filtering neon gas many thousands of times through pipe clay. The work was laborious and tedious but partially successful: Aston noted a change in the density of the filtered gas of 0.7 per cent.

After the First World War, in 1919 Aston constructed the first of three instruments to continue his research. The principal difference between the mass spectrograph and Thomson's positive ray apparatus is that the spectrograph uses both magnetic and electrical fields working together to progressively separate beams of particles of equal mass. The net result is that when the beams arrive at the detector end of the flight tube (in early instruments, this was invariably a photographic plate) they were sufficiently separated to be distinguishable as bands of differing intensity. This is, of course, the mass spectrum from which Aston's instrument took its name. Aston almost immediately used it to discover a third isotope of neon, mass 21, that proved that the mass spec (as it is now universally abbreviated) was indeed the technique of choice when measuring the differences between isotopes.

At the same time, Soddy's theory of isotopy was becoming imbued in the scientific consciousness; no longer a strange sub-phenomenon associated with radioactive materials, isotopy was a fundamental feature of the atomic world. Aston went on to describe 212 of the 287 naturally occurring isotopes with the aid of the machine that he invented. He also formulated the 'whole-number rule' which explains why atomic masses are often perplexing fractions of a whole number. Aston pointed out that these were merely averages and that in fact the atomic weight of nearly all isotopes is always a whole number (the exception is hydrogen). As atomic theory developed, the reason for this became clear. The differences in the masses of isotopes is due to varying numbers of neutrons. They have mass but no charge and therefore different isotopes of the same element do not have different chemical properties.

It was in 1929, the same year that Dubey and Holmes published their first successful helium-based dates on igneous rocks, that Aston – using the mass spectrographic technique – published the first measurements of the abundance of lead, the element ultimately produced by the decay of uranium. Rutherford, by now well-ensconced as the head of the Cavendish was not slow in using the new data to recalculate the still contentious age of the Earth.

But Aston's route to the analysis of lead was not as easy as his initial measurement of the inert gases, neon, argon and krypton. Since he developed the first instrument in 1919 he had been working on expanding its capabilities beyond merely identifying the *presence* of isotopes to estimating how *much* of each isotope was present by using the brightness and width of the lines in each sample's mass spectrum. His early attempts to quantify the isotopes of lead in this way by using a powder derived from ordinary metallic lead had not succeeded however. Another form of lead was needed. Aston had been working cooperatively and successfully with C.S. Piggott of the Geophysical Laboratory of the Carnegie Institution of Washington DC (*see* Chapter 8) and it was Piggott who in 1927 sent Aston a sample of pure lead tetramethyl which he had obtained from the US Chemical Warfare Service. (Nowadays, we are familiar with lead tetramethyl as the agent that is added to low-octane fuels to prevent 'knocking' in car engines. It is also a toxin that affects the central nervous system, causing symptoms ranging from mild insomnia to convulsions and coma, which presumably explains the Chemical Warfare Service's interest in it in the aftermath of the gas attacks of the First World War.) Aston's first measurements of this terrifying substance were much more successful and allowed him to calculate the relative abundances of the three isotopes of lead that it contained. He found the relative proportions of the isotopes Lead-206, Lead-207 and Lead-208 in the sample to be 4:3:7 which, when combined and averaged agreed well with the estimated atomic weight of lead of 207.2. Having thus established the isotopic composition of a standard, the next step was to repeat the experiment using radiogenic lead, which is lead derived from a mineral containing uranium. After a good deal of trouble, Piggott obtained a sample of the uranium-rich mineral uraninite from a dealer in Norway and sent it to his friends in the US Chemical Warfare Service to be converted into lead tetramethyl. A sealed tube of the resulting concentrate was sent to Aston early in 1928 but arrived broken! A second tube was prepared from the remaining specimen in Washington and sent to Aston that summer. It was found to contain a high proportion (that is it was highly enriched in) the lighter isotope of lead – Lead-206. The Earth's age calculated on this isotopic basis was 909 million years, a figure that compared favourably with that based on the *chemical* lead–uranium technique. But the data went even further, as clearly visible on the photographic plate from the Norwegian specimen was a line indicating the presence of the lead isotope Lead-207. Aston showed the results to Rutherford and both came to the same conclusion, the Lead-207 must have come from the only other known uranium-to-lead decay pathway, the so-called 'actinium' series: the ultimate parent of this series was a previously unknown isotope of uranium, Uranium-235. Knowing the decay constants of both Uranium-238 and Uranium-235 Rutherford was then able to compute, via two equations, the age of the Earth. His estimate was 3.4 billion years, in 1929 by far the oldest estimate of the age of our planet and one that is relatively close to our present-day estimate of over 4 billion years.

Aston's machine had truly revolutionised the science of real-time.

The Minnesotan

Far in the north of the United States, in the middle of the great plains and not far from the Canadian border, lies the enormous urban sprawl of the twin cities of Minneapolis–St Paul. I used to visit MSP (to give it its universal flight-destination code) regularly when I was doing a lot of science in the States in the early nineties. There was something vaguely surreal about flying across mile upon mile of open, seemingly uninhabited prairie and then, as the 747 banked and the spoilers sent you sliding down the drain, seeing the giant island of concrete with its central ring of skyscrapers reaching defiantly towards the endless sky. As soon as I landed I would head immediately for gate 13 at the far end of the giant concourse. It must have been a mile on the moving walkway but it was worth it. For gate 13 has the best bar in the entire airport. I'd often sit there, nursing a beer and staring out at the tarmac, thinking about the role that MSP played in developing the new science of fossils. One year, during a two-hour lay-over and with the snow sifting endlessly out of a leaden sky, I even made a pilgrimage into the city itself and just stood there, outside the main shopping mall – drinking in the bright Christmas lights and the residents' *bonhomie* – watching the manhole covers steam in the cryogenic air and thinking some more. For MSP was the home of Alfred Nier, the father of the type of *modern* mass spectrometer that we still use – virtually unchanged – today.

Al Nier was born in St Paul Minnesota in 1911. He was always interested in science, excelling at maths at school and building elaborate devices with his Meccano set. At high school his interests expanded and he spent a lot of time in the local library consulting issues of *The Boy Mechanic* and building radio circuits from designs it published. He graduated from high school in 1927, the same year that Aston successfully measured the ratio of the isotopes of lead. In 1928, Nier went to the University of Minnesota to read electrical engineering, and in 1933 received his Masters degree. He stayed on to work for a PhD in physics. In 1936, the same year that Nier received his PhD, physicists Rose and Stranathan pointed out that since there are *two* decay series of lead (one leading from Uranium-235, the other from Uranium-238), the ratios of the two lead end-products themselves will tend to increase with time so that a third pathway – lead–lead dating – is available to confirm calculations based on the uranium–lead series.

Nier had chosen as his PhD advisor John T. Tate, a big name in the world of American physics before the war. Apart from being an excellent lecturer, Tate had a feel for those areas of physics where the most exciting discoveries were waiting to be made and he directed the members of his large coterie of graduate students in those directions. Tate was an incredibly busy man. He was editor of the prestigious *Physical Review* as well as a founder member of the American Institute of Physics which involved frequent commutes (by train of course) between Minneapolis and New York City and meant lengthy absences from the lab. As a result both of this and Tate's own belief that his graduate students be able to 'get on with it', the young Nier was left very much to his own devices. This was unfortunate, because Nier spent the first few months of his PhD studies concentrating on a project that Tate, on one of his infrequent visits to Nier's lab, immediately dismissed as something 'which the General Electric Company had done years ago, but had never bothered to publish'.

Nier overcame his mortification and teamed up with John Williams, a young post-doctoral scientist who had come to work with Tate on the development of a new mass spectrometer, a variant of the mass spectrograph that did away with the need for photographic plates. However, after a few months, Williams' interest in this new project evaporated when he and Tate, lured by the promise of success in the still wide-open field of particle physics, decided to develop a nuclear accelerator for the production of synthetic isotopes. Williams later went on to use this expertise in the development of the atomic bomb during the Second World War. But in the meantime Nier was on his own. Using the manual skills that he had first acquired as a boy with a Meccano set, Nier began constructing the mass spectrometer himself. He built an instrument not dissimilar to the one that Aston was using in Cambridge (the flight tube curved through 180 degrees). Within a few months he had found a third, naturally occurring isotope of potassium – Potassium-40. A decade later, Nier would use this discovery to develop a dating technique based on the decay of potassium to argon, the so-called K–Ar technique.

After completing his PhD studies at Minnesota, Nier moved to Harvard for a two-year stint of post-doctoral study. This was his introduction to the role of isotopes in geology and dating rocks. At that time the change-over from chemical measurements of the abundance of uranium and lead to those based on the isotope approach was well underway. Central though to successfully determining the age of the mineral was the ability to distinguish lead which had been created by the decay of uranium or thorium from non-radiogenic ('ordinary') lead. This was critical because if ordinary lead were confused with radiogenic lead, the amount of lead in a sample would be overestimated and therefore the age would be overestimated, too. Traditionally, the three types of lead were distinguished on the basis of their varying atomic weights. But Nier, using the mass spectrometer he built at Harvard, showed that the isotopic composition of ordinary lead could vary significantly even when the atomic weight did not. It turned out that the apparent constancy of the atomic weight of 'ordinary' lead in a sample was coincidental and that this lead could contain significant portions of radiogenic lead. This meant that previous age determinations (based around an analysis of the amount of uranium, thorium, thorium-derived lead, uranium-derived lead and ordinary lead) were all likely to be too young. This then was the death-knell for chemically based separation techniques. After Nier's work, the only reliable way to identify the different types of lead was by using the mass spectrometer.

Nier immediately realised that the ability to identify the different isotopes of lead with a mass spectrometer allowed the lead–lead approach advocated by Rose and Stanathan to be considerably refined. When coupled with analyses of other isotope ratios in a sample, several different age estimates could be calculated. If all were in agreement then the overwhelming likelihood would be that the age estimate was accurate. Even more exciting was Nier's idea that if a lead-rich rock sample could be found that had virtually *no* uranium or thorium in it then this could be taken to represent the isotopic composition of lead at the time of the formation of the Earth, for there would be no radiogenic lead in it. With this critical sample of 'primaeval' lead, all the prerequisites for calculating the age of the Earth would fall into place. Nier soon found such a sample in rock that was then considered to be the oldest on Earth: from Ivigtut in east Greenland. He published his analyses of the isotopic composition of this 'primaeval' sample in the *Physical Review* during some of the darkest days of the Second World War, in July 1941.

Others were not slow to capitalise on Nier's new data; no fewer than three scientists – Gerling in the USSR, Houtermans in Germany and Holmes in the UK – all working independently, used Nier's results to calculate the age of the Earth. They came up with results which placed the age of the Earth between 3.23 and 3.35 billion years. However the entire procedure was dependent on the idea that the isotopic composition of primaeval lead was accurately known, i.e. that no part of that lead had been added to by the decay of uranium or thorium. And it was not long before it was realised that Nier's Ivigtut sample was not as primaeval as it had been supposed to be.

The geochronologist Claire Patterson, working at Cal Tech in 1953, realised that a better estimate of the isotopic composition of primaeval lead would be forthcoming from an extraterrestrial source – iron meteorites. By the mid-fifties the consensus was that these had been formed at the same time as the Earth by the coalescence of the same primordial material that made up the early solar system, but they were also known to have very low abundances of uranium and thorium. Patterson measured a meteorite sample from the Canyon Diablo crater in Arizona and found that the abundance of uranium was indeed very low. Over the next few years, he measured other samples from different localities and in 1965 was able to calculate an age for the Earth of 4.55 ± 0.07 billion years. Patterson was then able to confirm elegantly and simultaneously that iron meteorites and the Earth were indeed the same age and that his estimate for the age of the Earth was accurate.

But the technical innovations that had led to a date for the age of the Earth did not help much when it came to direct dating of fossils or fossil-bearing rock sequences, for it tended to be confined to certain types of igneous rocks; fossil-bearing sedimentary sequences were datable only if they happened to have the right kind of igneous rocks intruding into some part of the succession. The breakthrough, when it came, was again from Nier. He and his colleague Aldrich followed up a suggestion first made by C. F. von Weizacker, the son of the German under-secretary of state in the days immediately before the Second World War. Von Weizacker realised that the inert gas argon was so abundant in the atmosphere that it might be forming through the radioactive decay of another common element with a similar atomic mass: he suggested specifically that the capture of electrons by the isotope Potassium-40 (that had been discovered two years before by Nier using his new high-precision mass spectrometer) could account for the high abundances of argon in the atmosphere. Aldrich and Nier's experiment was to measure the ratio of the isotopes Argon-40 and Potassium-40 in ancient minerals. They showed that these minerals were enriched in the radiogenic isotope of argon (Argon-40), a result consistent with the decay of Potassium-40 over geological time. The way was open for another dating technique based on another decay series, only one based on more common elements. By 1958, the potassium–argon method had been found to give reliable dates from basaltic minerals as well as from minerals that were often associated with sedimentary deposits: glauconites. Basalts are occasionally found associated with sedimentary sequences and this, together with the newly developed ability to measure sedimentary glauconites, enabled direct dating of sedimentary rocks and the fossils that they contained. By the early 1960s the K/Ar technique (as it had become known) was being widely applied to sedimentary successions.

The last version of the universal timescale that Arthur Holmes had published had been in 1947 and that had been based on just five uranium–lead and thorium–lead dates. With the advances in nuclear technology that had occurred in the war, the fifties saw the establishment of many new isotope-dating laboratories that were focusing not just on the old lead-based techniques but also on the new technologies of rubidium–strontium and potassium–argon dating. With so much new data available, the time was obviously ripe for a recalculation of the geological timescale. In 1959 Holmes published a revision of his 1947 timescale based on about twenty radiometric dates for the Phanerozoic. The K/Ar technique proved to be more suitable for dating younger portions of the timescale (especially the Cenozoic) and so he incorporated many K/Ar tie-points into his new compilation.

Holmes' timescale of 1959 represents the crowning achievement of a career which began when the discovery of radioactivity was less than a decade old. After a lifetime in which he helped define the science of geochronology, a lifetime in which he never gave up hope of establishing a universal timescale, today's timescale is little altered from Holmes' 1959 version. Arthur Holmes died in 1965 – truly an architect of eternity.

Yet the scientific mind is capricious. After the initial excitement had died down, the limitations of the K/Ar technique began to attract attention, specifically the fact that it could only be applied to those sedimentary sequences that contained either intercalated basalts or glauconites and that even dates as accurate as those based on K/Ar contained statistical uncertainties which grew inexorably larger the older the sample; K/Ar dates were neither perfect nor universally applicable.

Something more was needed, something that would more surely bridge the gap between radioactive isotopes and fossils. Arthur Holmes' dream of a universal absolute timescale binding the worlds of isotopes and life together could not after all be realised without this vital missing ingredient.

That ingredient was magnetism.

The grail of magnetic time

Rock magnetism and changes in magnetic polarity had been observed as early as 1853 by the Italian scientist Melloni, who showed that the direction of magnetisation of some ancient lavas from Mt Vesuvius was the same as the direction of the Earth's magnetic field. Interest in the succeeding decades was sporadic; in 1906 Brunhes showed that lava fields in the Massif Central of France were magnetised antiparallel to the present-day orientation of the Earth's magnetic field and in 1929 the Japanese physicist Matuyama obtained the world's first magnetic stratigraphy by noting the changing direction of magnetisation in successive basalts in Japan and China. None of these early workers had any idea that later extension of their observations would result in the establishment of the geomagnetic polarity timescale (commonly abbreviated to the GPTS), a tool which revolutionised the recognition of time in the Earth sciences. The method came to prominence in the West in the 1960s, but it is clear that the technique had already been taken seriously for some years behind the iron curtain; in 1957 A. N. Kramov had published a paper outlining the potential of the technique for correlating unfossiliferous strata – particularly of the Soviet Union's giant oilfields – in the *Proceedings of the Soviet Academy of Sciences*, a journal that was not widely available in the West.

The palaeomagnetic technique depends on the fact that magnetic iron oxide particles present in any newly deposited igneous or sedimentary rock align themselves with the prevailing direction of the Earth's magnetic field. This property is known as the rock's 'remanent' magnetism. Equally crucial is the fact that the earth's magnetic field has reversed polarity at frequent intervals in the geological past. The breakthrough that joined palaeomagnetic changes with the recognition of absolute time occurred in 1959 when Rutten and Wensink correlated volcanic lavas in Iceland with volcanic outcrops in France. They dated the lavas using the newly refined K/Ar technique, found that they did appear to be of the same age, and proposed that the boundary between the Pliocene and the Pleistocene was equivalent to a polarity change in the Earth's magnetic field. Rutten and Wensink thereby formalised an important concept that had first been put forward in 1926, namely that palaeomagnetic reversals were globally instantaneous. They also established an important precedent, which was that the sequence of palaeomagnetic reversals needed to be tied into absolute time using one of the radiogenic techniques.

Rutten and Wensink's success immediately stimulated two groups of young researchers to try to find other magnetic reversals in the rock record and tie them into Holmes' absolute timescale. Both were aware of the size of the prize; a single good K/Ar date was now all that was required to bind a globally recognisable and unique magnetic reversal irrevocably into the universal timescale. One K/Ar date would now do the work of dozens. The chase for the holy grail of magnetic time was underway.

The two competing groups were based in northern California and Australia. The US group comprised three young researchers – Cox, Doell and Dalrymple – all of whom had graduated from the University of California at Berkeley and later moved to government scientist positions at the US Geological Survey facility at Menlo Park not far from San Francisco. As graduate students they were exposed to the energy and enthusiasm of Jack Everdnen, in 1963 quite clearly one of the up-and-coming young men of geochronology. Everdnen was working on techniques for reducing the error associated with K/Ar dates and had made some spectacular advances, particularly in samples less than five million years old. There was even talk that the K/Ar technique would eventually overlap the age range of Willard Libby's newly developed technique of Carbon-14 dating and that there would then be no part of the palaeontological record that would not be accessible, however indirectly, to absolute dating techniques. Dalrymple was the expert on geochronology who performed the K/Ar measurements; Cox and Doell had complementary palaeomagnetic expertise. The group used Everdnen's laboratory to date six volcanic flows whose direction of magnetisation they had measured. The early results were promising – they suggested that the Earth might have a regular periodicity in its magnetic dynamo of one million years.

The Australian group was made up of McDougall, Tarling and Chamalaun at the Australian National University in Canberra and in 1963 they too published parallel palaeomagnetic and K/Ar data (from the Pacific island of Hawaii). The competition – by and large friendly – between the two groups led to the establishment within only five years of a geomagnetic timescale for the uppermost portion of the Cenozoic, the Pliocene and Pleistocene.

In the mid-sixties, there was no particular interest in dating sediments directly, although there had been some efforts in dating the basaltic underpinnings of the sea floor, on the assumption that the immediately overlying sediment would therefore be the same age. However, the argon necessary to date the rock had long ago been lost from its crystalline matrix and the attempt was abandoned.

The huge potential for accurate magnetic dating of sediments arose almost by accident as a by-product of an experiment to test the idea of sea-floor spreading, one of the major underpinnings of the then new theory of plate tectonics. In 1963 Fred Vine and Tujo Wilson published a paper in *Science* which showed that the pattern of magnetic stripes (or lineations) on the sea floor correlated with the magnetic polarity timescale as logged in basaltic rock sections by the American and Australian groups. As new rock is formed at the mid-ocean ridges, it incorporates the prevailing direction of magnetisation, and then moves away towards the ocean margins prior (many millions of years later) to sliding under the continents (a process known as subduction). Hence the pattern of magnetic stripes across the sea floor on either side of the mid-ocean ridges records the same pattern of polarity change that is found down-section in sedimentary outcrops and cores. Since the rate of sea-floor spreading appeared to be constant, one needed only to identify downcore magnetic reversals and correlate them with their equivalent on the sea floor to know precisely the age of any intervening sediment in the core.

This was a revolution: it meant quite simply that it was no longer necessary to date every magnetic reversal to calculate its age. Since sea-floor spreading proceeded at a constant rate (a few centimetres per thousand years), one had only to measure the distance of a magnetic stripe from the ridge axis to work out its age. Identify the stripe's equivalent downcore or in outcrop and, hey presto, you had an age for the rock and any fossils that it contained. By 1968 a group of scientists led by the geophysicist John Heirtzler had established a geomagnetic polarity timescale that extended all the way back to 70 million years before present (predating the beginning of the Cenozoic).

In the years since, the geomagnetic polarity timescale has been refined and extended much further back, as far as the top of the Jurassic. It is possible to date sediments and their fossils to within a few thousand years. Couplets of normal and reversed polarity are now called anomalies and have been assigned numbers, so that all of Cenozoic time and much of the Mesozoic can now be catalogued according to what anomaly it is associated with.

Of course there were – and still are – problems with the technique, the most significant being that the reversals have no unique signature of their own; they are either normal (shown by convention on stratigraphic logs as black boxes), or they are reversed (shown as white boxes). But fossils and fossil assemblages on the other hand *are* unique, and associating the anomalies with their unique fossil fauna has been led by analysis of deep-sea cores, with their characteristic fauna of planktonic foraminifera; Bill Berggren has spearheaded much of that research effort. The quest to define narrower and narrower bands of time within the accurately dated anomaly framework led to another group of fossils becoming widely used biostratigraphic indicators: the nannofossils. (Huxley's coccoliths are one of their two major constituent groups.) These tiny fragments of calcium carbonate are the remains of the free-floating algae on which the planktonic foraminifera themselves fed and yet they are a hundred times smaller even than the planktonic foraminifera. However, nannofossils evolve and their populations change their characteristics more quickly than the planktonics and so they are even more useful for recognising smaller and smaller chunks of time: in some cases intervals of less than 100,000 years in sediments that are more than 60 million years old.

So the age of real-time in the new science of fossils had arrived. Isotopic-dating techniques provide the tie-points for the geomagnetic polarity timescale which works as readily in sediments as it does in basalts, and the recognition of the anomalies is reliant on their unique fossil content. As with relative time, the science of real-time could not operate without the new science of fossils.

Weighing Oceans

A Glass of Heavy Water. Technische Hochschule, Zurich, Switzerland, 47.15N, 08.18E. 10 December 1946

The figure at the podium was spare to the point of gauntness. As he leaned over it, the room seemed filled with his presence. Harold Urey's dark eyebrows above his smooth and still boyish face were a continuous line across a forehead furrowed in thought. He was struggling to persuade a lay audience, who had accepted an idea that had been radical enough to upset the fundamentals of physical chemistry only thirty years before, that now they would have to upset their thinking all over again. He was trying to tell them that different isotopes *did* have different properties after all. As he looked across the room, he realised that this august grouping, the best and the brightest scientists in middle Europe in this, the aftermath of the bloodiest conflict that the world had ever known, believed to a man the isotopic doctrine that Soddy had sanctified three decades earlier: isotopes of the same element, whether produced by radioactivity or not, *did not have different chemical properties*. They only weighed differently. Their world view was as unshakeable as the atomic structure of the nuclei of the stable isotopes themselves. It would not easily be persuaded to change.

Eventually he held up the tumbler of water that had been beside him throughout his lecture. He asked them to imagine the different water molecules within the glass. The molecules, he reminded them, were of linked hydrogen and oxygen atoms: two hydrogen and one oxygen combining to make up the molecule H_2O. But the various molecules of water – although chemically identical – did not have the same atomic weight. Although the oxygen atom was on average about sixteen times heavier than the hydrogen atom this was only an approximation. The 'about' came in because oxygen had three isotopes, oxygen-16, oxygen-17 and oxygen-18 and it was the contribution of the uncommon oxygen-18 and the very rare oxygen-17 that, together with the common oxygen-16, gave the element oxygen its *average* atomic weight of a little over 16. Hydrogen, he reminded them, also had two naturally occurring isotopes, hydrogen itself and its heavier cousin, deuterium, but these were so 'light' the difference between them was negligible.

The laureate stopped for a moment, his gaze far away as he remembered that day in 1933 when he himself had been awarded the Nobel prize for the discovery of that second hydrogen isotope. And of that day's other prize, the birth of the son which had prevented his attendance at Stockholm to receive his prize. Then he reached the crux of the matter: although isotopes were chemically identical, he explained, the tiny differences in mass *did* give rise to small differences in reactivity. This wasn't anything to do with the *chemical* properties of the isotope; that, as Rutherford and Soddy had shown, was a function of the number of orbiting electrons. No, this was to do with simple atomic weight. He placed the glass with deliberation on the table. If he waited long enough, he said, the glass would get heavier, as the water inside got heavier. The audience was silent, trying with difficulty to grasp the concept. Why would the water get heavier? Because, Urey said, those molecules of water with the light isotope of oxygen – oxygen-16 – would evaporate more readily than the other type of molecule with an *oxygen-18 atom*. The molecular bonds that held it to its neighbours would break more easily than water molecules containing the heavier oxygen-17 or oxygen-18. The audience stared at the tumbler, trying to see it settling more and more heavily into the chalk dust on the bench. The water, Urey concluded, was getting heavier as the oxygen-16-carrying water molecules evaporated more readily from it. The water left behind was becoming enriched in the heavier isotopes; particularly oxygen-18. And thus was the glass of water itself becoming heavier.

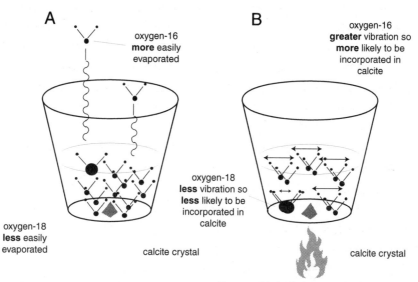

Fig. 4.1. The two controls on the distribution of oxygen isotopes. A. The glass of water originally envisaged by Harold Urey. Molecules of water containing the common, light isotope of oxygen evaporate more readily than molecules of water containing the uncommon, heavy isotope of oxygen. B. When the same glass of water is heated the molecules of water with the light isotope of oxygen vibrate more rapidly than those containing the heavy isotope of oxygen. Mineral crystals growing in this solution reflect these effects.

The silence in the room was itself heavy as Harold Urey's audience – the great and the good – come to hear this famous man, absorbed this ridiculous piece of nonsense.

'But there is more to this than that.' The laureate stared across the room. 'If the water in a tumbler can increase in weight during the course of this lecture – one single hour – think what must have happened to the oceans in the course of geological time.' There was a silence and then a stir from the audience as a man in the front row hesitantly put up his hand.

'The isotopic composition of sea water will become enriched in the two heavier isotopes of oxygen as the light isotope, the oxygen-16, evaporates back into the atmosphere?'

Urey nodded. He recognised the speaker, Paul Niggli, one of the brightest chemists in ETH, a man who specialised in the chemical properties of crystals. There was a sudden tension in the air, as though something of fundamental importance, some insight that the laureate had missed was suddenly materialising in the lecture theatre in front of him. The young man continued thoughtfully, 'But surely it will be possible to measure that? Any mineral precipitated in that water will reflect its isotopic composition.'

There was a silence and then the laureate said thoughtfully, 'That's true too.'

From so simple a beginning was one of the most important techniques in the new science of fossils born . . .

Harold Urey was one of the most brilliant chemists of the twentieth century. His name is also linked with Stanley Miller as together they proved that organic compounds could be forged from inorganic materials in a sealed container, thereby proving that life could have originated spontaneously in the primordial oceans. Urey was born in 1893 in, as they say, the buckle of the bible belt: Walkerton, Indiana. He was the son of a preacher and a devoutly religious mother, themselves children of pioneers who had settled in the territory. By the time he was eighteen, and with only a country education to his name, he was already teaching in high schools. In 1914 he entered the University of Montana where he read zoology. After spending a couple of years as an industrial research chemist, he returned to Montana briefly before moving to California where he was awarded his PhD in 1923. Urey was a type of scientist whom you rarely encounter today. He was a scientist who had an eclectic interest in the nooks and crannies of his subject and the means and temperament to explore widely. These days such unfettered intellectual exploration is condemned as 'blue-sky research' and does not attract funding. But in Urey's day and age it was permissible, and to mankind's lasting benefit this is precisely what Urey did.

Urey's crowning achievement was the development of the oxygen isotope thermometer. After the Swiss meeting Urey returned to Chicago where he did a few back-of-the-envelope calculations. And he discovered that if the ocean were monitored by proxy (for example, by the hard parts of sea-dwelling creatures as Niggli had suggested), then *another* factor would become important too: temperature. The proportion of the isotopes of oxygen incorporated into the shell of an animal that had lived in that ocean would vary not *only* as a function of evaporation, it would also vary with the temperature of the water – the higher the temperature the more the light oxygen – oxygen-16 – would tend to become incorporated into the animal's shell. When measured as a ratio of oxygen-18 to oxygen-16 (oxygen-17 was so rare it could be effectively ignored) the warmer the water then the more that ratio would decrease as more oxygen-16 became incorporated.

'All of a sudden,' Urey said, 'I found that I was holding in my hands a geological thermometer.'

The reason that oxygen-16 and oxygen-18 are the two isotopes of interest to the palaeothermometrist is simply one of relative abundance: oxygen-16 accounts for well over 99 per cent of all the oxygen known on the planet, oxygen-18 accounts for a much, much smaller proportion; about 0.5 per cent. Urey assumed that the proportion of oxygen-16 to oxygen-18 in sea water was approximately 500–1. But factoring in the influence of temperature on the way that the different oxygen isotopes would be incorporated into chalk shell secreting organisms (or indeed even chalks that were not precipitated by a biological agency, so-called 'inorganic carbonates') Urey calculated that the ratio will be very slightly offset from the correct ratio of 500-1. He found that at 0°C the ratio of oxygen-16 to oxygen-18 would be closer to 500-1.026, or to put it another way, at a temperature as low as the freezing point of water one additional oxygen-18 atom would be preferentially incorporated into a calcium carbonate lattice for every 19, 231 oxygen-16 atoms. Similarly, at a temperature of 25°C (a temperature close to the maximum found in the warmest parts of the world's oceans) he calculated that less of the heavy isotope of oxygen would be incorporated; in this case one atom of oxygen-18 would be incorporated for every 22,727 atoms of oxygen-16. In other words for every one degree rise in temperature 140 *fewer* oxygen-18 atoms are incorporated.

Herein lies one of the main problems that confronted Urey: just exactly how do you go about measuring the differences in atomic weight conferred by at most a few thousand atoms?

Before we see how Urey and his team solved this problem let's just take a moment to consider the subtlety of Urey's insight: 140 atoms! When one considers that the oxygen isotope method is the technique most widely used today for measuring temperatures in the geological past, the fact that it was based on a theoretical calculation of the differential distribution of a handful of atoms (calculated on the back of an envelope in the immediate aftermath of the Second World War) makes Urey's achievement one of the crowning insights of the new science of fossils.

But the problem was turning theory into practice. There were three problems that needed tackling: first was to improve the precision of existing mass spectrometers so that they could measure the tiny variations in atomic weight required; second was the problem of turning the chalk sample into a gas where the isotopic abundances could be measured; and third was cross-checking results with theory to be sure that the technique actually worked. The final step, of converting the oxygen isotope ratios to actual temperatures, was the easy part. That could be handled by some basic algebra!

The Chicago Mafia

Urey realised pretty soon that he was on to something quite remarkably important. But by the time he was fifty, Urey, like many of the best scientists before and since, had simply recognised that he couldn't do it all. He knew that if he wanted the technique to work he would have to let students do the work that would result in the development of the oxygen isotope thermometer. For any scientist, this is one of the hardest of all decisions to make. Not only will these young Turks do the work which you no longer have time to do yourself, but they will also take credit for the necessary discoveries that will turn your beautiful theory into practice.

Among the first young men that Urey recruited into the Chicago Mafia was John McCrea. If Urey was the theoretical genius behind oxygen isotope palaeothermometry, John McCrea is arguably its technical founder. Yet when he arrived in Chicago in the late 1940s he was fresh out of college, and had come to work with Harold Urey in order to obtain his PhD. McCrea was given the awesome responsibility of developing techniques that would turn solid calcium carbonate samples reproducibly into oxygen atoms that the mass spectrometer could then count. This meant taking Nier's mass spectrometer and modifying it to work on small gas samples. (By this time Nier, that tireless Minnesotan engineering genius, had developed instruments with flight tubes that curved through only 60 degrees. The 180 degree instrument that had been developed by Aston and then modified by Nier had become obsolete.) The technical problems, as Urey knew well, divided broadly into the chemical and the instrumental, and so while McCrea was given the former, his young colleague Charles McKinney was given the latter.

For McCrea it was clear that the geological materials that the Chicago Mafia would try to analyse first would have to be calcium carbonates. This was partly because this mineral is the most abundant form of fossil material on the planet, partly because it is readily soluble in acid, and partly because Urey had devised the theoretical basis of the oxygen isotope thermometer on the basis of carbonates. (I should mention here that there are more forms of carbonate than just *calcium* carbonate; see the Glossary for a more detailed explanation.)

McCrea settled down to think about how to turn carbonates into gas. There were effectively only two approaches: you either heated it so violently it broke down into gas, or you digested it using acids. It was clear early on that the thermal decomposition method would not work as the results were far too variable. The digestion method seemed a simple proposition: just pour sufficient quantities of acid on to a carbonate sample and it would turn into gas. The problem though was how to do this reproducibly (over and over), digesting sub-samples from the same specimen and then getting the same result. Over the next few months McCrea experimented with different acids. Hydrochloric and sulphuric acids would not work properly because they contained too much water. Water, H_2O, has a great deal of oxygen in it and tends to want to exchange with the carbon dioxide being produced by the reaction, and any water that gets into the mass spectrometer will itself contribute to the ion beams that carry the oxygen atoms needed to reconstruct ancient temperatures. It is a contaminant. Phosphoric acid was better, but it was clear that there would always have to be *some* contamination from water. For water is a by-product of the very reaction which turns calcium carbonate into the gas carbon dioxide. Calcium carbonate when reacted with phosphoric acid yields three daughter products; ordinary water (H_2O), the mineral calcium phosphate ($CaPO_4$), and carbon dioxide (CO_2). It is only the CO_2 that is required for the measurement – for this carbon dioxide contains the same proportions of oxygen atoms as the calcium carbonate sample. The other two reaction products need to be separated from the gas to be measured. The calcium phosphate is not a particular problem, it stays dissolved in the acid, but the water *is* a problem. McCrea was a consummate chemist though, and saw the answer immediately: a device called the cold trap. As the water vapour/carbon dioxide gas mixture passes into the pipework above the mass spec it is made to pass through a region of pipe where the temperature is maintained at $-196°C$. This is cold enough to freeze out both the water and the carbon dioxide but not any contaminants. After the gas is safely frozen, the region above it is pumped clean. The trap is then warmed to $-100°C$, warm enough to free the H_2O but not warm enough to free the CO_2. So in one step you strip the gas mixture of its major contaminant and no complicated chemical jiggery-pokery is required at all. The CO_2 containing the oxygen atoms is then free to make its way onward towards the mass spectrometer itself. And all of this is merely the prelude to the actual analysis, the trailer, as it were, for the main event. And here was the second problem, how to get the gas into the machine and then make a meaningful estimate of the number of oxygen atoms in it. And that was McKinney's problem.

The mass spectrometer that McKinney decided to use was Nier's 60 degree instrument. It was the most commonly available and, by 1950 had been proving its worth for over ten years. It also conveniently lent itself

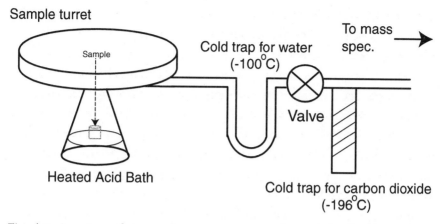

Fig. 4.2. A variety of preparation system commonly used for the liberation of carbon dioxide gas from carbonate samples. This 'bucket chemistry' is an automated version of the technique developed by John McCrea in the early 1950s. The only thing that has changed is that a computer now oversees the reaction, handling and analysis of the sample.

to the analysis of either solid samples – condensed as a salt (lead for example) on a filament that was then heated – or gaseous samples. The Nier instrument though was not capable of measuring isotopic abundances to the degree of accuracy that Urey had calculated would be required for palaeotemperature work. Urey had estimated that a change in temperature of one degree centigrade would result in a change in the proportions of oxygen-18 to oxygen-16 of only two hundredths of one per cent! The Nier instrument – the best mass spectrometer then available – was only capable of measuring differences to a fifth of one per cent, equivalent to 10°C. A tenfold increase in the sensitivity of the instrument was required. It was a tall order, but McKinney grasped the nettle and set about modifying the instrument to enable it to measure smaller differences. He very soon realised that low abundance isotopes like oxygen-18 could not be measured in *absolute* terms with any degree of precision. They would have to be measured in relative terms. It would be more accurate to compare the proportions of the two isotopes in the unknown sample with the pro- portions in a gas of *known* isotopic composition. And thus was the isotope *ratio* mass spectrometer born. McKinney devised a sequence of valves and pipework that was – until very recently – still the defining characteristic of those mass specs that are used to weigh the atoms of fossils. The essence of this dual inlet is that it is symmetrical – the valves and pipework necessary to move the gas generated by the sample are mirrored by the valves and pipework needed to move a 'standard gas' (that is of already known isotopic composition) from a tank to the mass spec. Coupled to this McKinney used a special type of valve to admit the gas into the ionisation chamber of the mass spectrometer. Although this has always been known as a McKinney valve it was in fact designed by one of Nier's co-workers in the University of Minnesota, R. E. Murphey, who had become involved in the race to improve Nier's instrument for stable isotope work. The function of the Murphey valve is to allow multiple estimates of the isotopic composition of the standard and the sample gas. In this way an average of the isotope differences can be taken, vastly improving the precision of the final result. It was a brilliant idea, and one that dramatically improved the accuracy of Nier's mass spectrometer.

McKinney's final technical innovation was to improve the sensitivity of the amplifiers in the electronic circuitry of the machine so that the innovations in the gas-handling system could be fully exploited. By the early years of the 1950s these technical innovations were 'on line' and Urey's group was finally able to address the problem of measuring the temperature of ancient oceans.

The result of this enormously hard and tedious technical work was the development of a simple equation that took the ratio of oxygen isotopes in a calcium carbonate sample and converted them to temperature. Urey's back-of-the-envelope calculations pointed the way but there was no substitute for testing the relationship in practice or, as scientists like to say, calibrating it empirically. The Scripps Institution of Oceanography on the shores of the Pacific Ocean in La Jolla provided calcium carbonate samples from molluscs and other shelly creatures that had been grown in tanks at controlled temperatures as well as other calcium carbonates that had been inorganically precipitated. McCrea worked on these while another student of Urey's, Samuel Epstein, worked on the carbonates from wild marine creatures. He concentrated particularly on the various kinds of molluscs as well as brachiopods (stalked mollusc-like creatures, now very rare, that once dominated the oceans of the Palaeozoic). Early results from these groups suggested that their oxygen isotope proportions accurately reflected that of the surrounding ocean water. Other groups, such as echinoderms, were more problematic and suggested that their metabolisms interfered with the isotopic proportions as the shell was grown. Very soon this problem was given the name 'vital effect'. The Chicago Mafia also established a standard against which all results were to be reported: a fossil called a belemnite from the Pee Dee rock formation in Carolina. This standard scale for reporting oxygen isotope values is still in use today. It is called, with a staggering lack of imagination, the PDB scale.

The view from the Fen

By the late 1950s, the ability to measure the temperatures of fossil oceans was well established and the technology was filtering out from Kent Nuclear Laboratory in Chicago as one-by-one Urey's young disciples left the nest and set up on their own. The young Mafiosi found themselves in considerable demand as the implications and scope of the palaeothermo-metry technique became apparent to faculty chairmen across America. Perhaps unsurprisingly, one of the first of the new labs was established at the California Institute of Technology which successfully recruited Sam Epstein early on. For three decades the Caltech group set up by Epstein remained one of the most vibrant isotope labs in the world. But it was Cesare Emiliani, another of Urey's Mafiosi, who early on decided to concentrate on the planktonic foraminifera and took a faculty position in Florida to give himself time to develop the technique, who would become the king of the palaeotemperature world. Soon the excitement of the isotope thermometer was spreading overseas. An Italian named Longinelli set up a lab in Trieste while Eric Olausson established a facility out on the periphery of Europe at Gothenburg. And in Cambridge a visionary named Harry Godwin thought that Britain was in danger of being left behind. Godwin was the University's Professor of Botany but he was also the head of an institute within the university devoted to the study of the Quaternary. (The rock outcrops across East Anglia are often of these very recent sediments and Cambridge took advantage of their proximity.) Godwin was a man of disarming energy and enthusiasm who was also fascinated by technology. It was he who bought the department its first electron microscope and as the first grainy image of a pollen cell flickered on to the screen could not restrain a little boy's breathless, 'Cor, look at that!'

As news of the oxygen isotope technique spread across Europe and the importance of Emiliani's new paper *Pleistocene temperatures* became apparent Godwin knew that Cambridge would have to enter the fray or risk losing its position at the cutting edge of Quaternary research. Godwin cast around for someone who could get the enterprise up and running and eventually selected a young Cambridge undergraduate named Nick Shackleton fresh from the natural science tripos to set up the laboratory while pursuing research for his PhD degree.

This young man was thrown in at the deep end. Nick Shackleton struggled to enter a new science dominated by the slight, distant figure of Cesare Emiliani. Shackleton once remarked to me that for the first few years after he'd started his research, every new idea that he came up with had already been anticipated by Emiliani. But after a while Shackleton recognised that Emiliani was approaching the problem of glaciations in rather too straightforward a manner. The more he thought about it, the more convinced he became of a fundamental flaw in Emiliani's interpretation of the oxygen isotope record.

Urey's original concept was that of the ocean's gradual enrichment in the heavy isotope of oxygen (oxygen-18) as the common light isotope (oxygen-16) more rapidly evaporated back into the atmosphere. Urey's back-of-the-envelope calculations had then revealed a genuine, no-nonsense, 24-carat geo-thermometer. In the years after that, McCrea, Epstein, Emiliani and the other members of the Chicago Mafia had naturally tended to interpret the ancient oxygen isotope record more as a temperature record than anything else. In 1955 Emiliani had calculated that 80 per cent of the change between glacial intervals and interglacials was due to temperature and only one-fifth was due to variations in the ratio of oxygen isotopes in the ocean because of the growth and decay of ice-sheets. This was Nick Shackleton's breakthrough. He realised pretty quickly that Emiliani was making a big assumption. It was already known that the isotopic composition of the ice near the Pole of Inaccessibility in Antarctica was −60 parts per thousand (per mil), or four times *more* deficient in oxygen-16 than Emiliani had estimated that ice caps *should be*. The disparity was of more than academic interest; for although increasing temperature means that more of the light isotope of oxygen gets incorporated into a carbonate mineral, the end of a glaciation will have the same effect: as oxygen-16 is returned to the ocean as the ice caps decay more oxygen-16 becomes available to marine organisms to build their shells from.

Shackleton decided that he would devote his graduate work to a reassessment of Emiliani's glacial guesstimate. He thought long and hard about how to approach the problem before a solution dawned on him. And the answer lay not with those foraminifera which lived in the waters of the surface ocean, but with those that lived on the sea bed, the so-called 'benthic' foraminifera.

The essence of Shackleton's idea lay in the fact that deep-water temperatures were supposedly unchanging. Deep waters today are only a degree or two above freezing and therefore cannot cool much further when the world enters a glacial period. He would measure the oxygen isotope ratios of benthic foraminifera (that lived on the sea bed), then he would compare them with the ratios in planktonic foraminifera (that lived in the surface ocean) and from that difference calculate the magnitude of the ice-volume effect versus the temperature effect. But the technical problems were once again formidable. Benthic foraminifera are scarce. So either more must be found to provide enough CO_2 for the mass spectrometer to analyse, or the sensitivity of the machine to this tiny amount of gas must somehow be further increased – or both.

Cambridge had no isotope lab like Chicago and La Jolla and no one like McKinney whose skills were devoted to maintaining the instrumentation. Shackleton had to do it all. With the aid of a grant from the Department of Scientific and Industrial Research, a primitive mass spectrometer was purchased. He added the features that McCrea and

McKinney had invented a decade before and then began the laborious task of modifying still further the inlet and electronics systems to measure truly tiny quantities of carbon dioxide gas. After a couple of years of painstaking development work he was finally in a position to measure the handful of specimens found in deep-sea core samples and address the question of the relative contributions of ice growth and temperature change in the Pleistocene glacial ages.

In the end, Shackleton only succeeded in making a few measurements for inclusion in his thesis. But those measurements were enough for him to recalculate Emiliani's estimates of the relative contributions of ice volume versus temperature change. And Shackleton showed that the changes in benthic oxygen-18 were virtually as large as those in the planktonic foraminifera and therefore that Emiliani had to be mistaken. Almost the entire oxygen isotope signal came from variations in the isotopic composition of sea water. Almost none came from Urey's temperature effect after all.

The ice machine

Shackleton's reputation was assured. But in Brown University in Rhode Island another academic was wondering about the ice ages and his emphasis was not on the effects of ice-sheet growth and decay but rather on the causes. His name was John Imbrie and his preoccupation was with the theories of an obscure early twentieth-century Czech astronomer named Milutin Milankovitch. Milankovitch had hypothesised that the causes of the ice ages could be traced to variations in the way the earth orbited around the sun. The idea was of brilliant and almost unbearable simplicity. Milankovitch had calculated that the variations in the sun's energy falling on the surface of the earth as the elements of the earth's orbit varied were the causes of the temperature variations that gave rise to the ice ages.

For close on four decades this idea knocked around the nooks and crannies of pre-Nazi Europe. In the white heat of that conflagration, the ideas were partially forgotten but somehow resurfaced in the aftermath of the war. And John Imbrie was their champion. By the middle sixties Imbrie had abandoned a classical palaeontological career and had devoted his life to testing Milankovitch's hypothesis. He heard of the new technique of oxygen isotope palaeothermometry and in particular of Shackleton's reassessment of Emiliani's original theory. One of the consequences of Shackleton's emphasis on the evaporation/precipitation control on oxygen-isotope ratios was that the sequence of glacial ages as seen in the oxygen isotope record of deep-sea cores was now freed of the complicating effects of temperature, and could be seen as accurately portraying the rate, amount and *timing* of the growth and decay of ice sheets. Imbrie realised that this was the technique that he had been waiting for. The hunt was on for the right deep-sea core. The trick would be in getting a succession of uninterrupted oxygen isotope stages (that is glacial–interglacial cycles) in one single core. Such a core had been drilled in the Caribbean, and Shackleton and Imbrie, now enthusiastic collaborators, in conjunction with the geologist John Hays succeeded in extending the oxygen isotope record back through several glaciations or oxygen isotope stages.

Imbrie then produced the crucial additional ingredient that would connect Milankovitch's orbital cycles with Shackleton's oxygen isotope stages: the maths. This technique, known as spectral analysis, in the 1970s was very new. Effectively it summarises in a simple graph the complex information contained in the varying wavelengths of a signal – an analogue radio wave for example. This explains the Imbrie group's obsession that the core should not contain any gaps. For if the record is complete, then the oxygen isotope variations from a core can be treated just like a radio wave.

Milankovitch's hypothesis identifies three components to the Earth's orbit as it travels around the sun. The first is the eccentricity component: the track of our planet's orbit around the sun is neither perfectly circular nor is it a conveniently invariant ellipse; it varies, stretches, with a cycle of something like 100,000 years. The second is that Earth's axis is also tilted with respect to its orbit around the sun and this varies with a period of about 40,000 years. Finally, the Earth's axis wobbles like a child's spinning top skating across a rough concrete surface and this has a frequency of about 23,000 years. These components interact to control climate, making the oxygen isotope record one of apparently intractable complexity.

The working hypothesis of the Imbrie team was that all three components would be encoded into the oxygen isotope record and separable by spectral analysis. Hays, Imbrie and Shackleton applied the technique to their core and found that three peaks resolved themselves: one at 100,000, one at 42,000, and the last at 21,000 years. Milankovitch was vindicated. The succession of the Pleistocene ice ages was indeed controlled by variations in the Earth's orbital geometry.

The carbon connection

By the early 1970s mass spectrometers for stable isotope work were becoming more common as the potential of the new science of oxygen isotope palaeoclimatology became recognised by more and more universities across the world. The earliest 'off-the-shelf' instruments were not sold with the preparation-system needed to turn a carbonate sample into carbon dioxide gas. These still needed to be manufactured in-house and many a departmental workshop was soon turning out contraptions of varying Heath Robinson complexity to do precisely this job. A consequence of the semi-unique nature of each system was that it was still necessary for the isotope lab to be run by experienced and dedicated technicians. Shackleton recognised this early on and in perhaps his happiest stroke of good fortune teamed up with a young man named Mike Hall. The two of them pushed back the technological boundaries of the technique, progressively modifying the machines until they could run samples as small as only seven individual forams – less than a tenth of a milligram in weight.

The gas that these machines measured was CO_2, but the carbon component of the gas was almost always ignored in routine palaeoceanographic work. But in the late seventies carbon too was seen to be involved in climate control through its role as a greenhouse gas. It was soon realised that the ratio of carbon isotopes was related to the rate of productivity in the oceans. (The implications of the carbon system will be discussed in more detail in Chapters 6 and 7 but for now it is enough to know that 'productivity' is technically defined as the rate at which carbon is fixed into plant tissue via photosynthesis. And it is photosynthesis – rather than temperature or the growth of ice sheets – that controls the incorporation of carbon isotopes into living tissue and calcium carbonate. The photosynthetic pathways of all green plants prefer the lighter isotope of carbon (carbon-12) for much the same reasons that a glass of water loses oxygen-16 faster than oxygen-18: its lightness makes it more mobile.)

The ability to measure the rate at which photosynthesis fixes carbon (whether in oceans or in more enclosed water bodies such as lakes) is the entrée into several important new fields of the new science of fossils. In order to detect this other isotope ratio (carbon-13 to carbon-12) the mass spectrometers needed modification.

The big 'D'

In the heady days of the fifties and sixties the feeling running through the global palaeontological community was that the new oxygen isotope technique – and the even newer carbon isotope technique – was the Philosopher's Stone. They had done it! They had unearthed the alchemy that would grant them the secrets of Earth's climatic record. But, as is so often the case, lurking just around the corner reality was waiting to smack the new palaeontologists in the eye. It turned out that nature was no more likely to give you a free lunch than your own head of department. The problems with the stable isotopic technique began to come out of the woodwork.

There are two main reasons why the technique of weighing atoms is not as straightforward as one would like. The first of these is that we do not know for sure that the ratio of oxygen and carbon isotopes that a foram (or some other animal) incorporates into its shell accurately reflects the ratio that we need to measure – which is the ratio that exists in sea or lake water. Allan Bé, a biologist working at the Scripps Institution of Oceanography in La Jolla, California, saw that an entire discipline was being founded upon the biology of a very poorly understood group of single-celled organisms. It was Bé who pioneered the culturing techniques that allowed planktonic foraminifera to be grown in tanks. After Bé's untimely death in the early eighties Jonathan Erez, an Israeli biologist working in the Gulf of Elat and Howie Spero at the University of California, spent the next decades trying to understand whether planktonic forams really *do* accurately reflect the ratios of oxygen and carbon isotopes in ancient sea water. Fortunately for the new science of fossils Erez showed that most planktonic forams do accurately reflect the oxygen isotopic composition of the water that they grow in. However, it was Spero who showed that carbon isotopes are often influenced by the metabolism of their host animals. At first this appeared to discredit the carbon isotope theory, but not for long.

The second complication is that stable isotopes are susceptible to alteration *after* they have been incorporated into a fossil shell (or inorganic carbonate cement). This whole knotty problem is known as 'diagenesis' – what I call the big 'D'. The problem occurs because the isotopes of oxygen and carbon are relatively easy to move around. It is hard to see how it could be otherwise – it is this very property which allows their proportions to vary and hence, when incorporated in fossils, makes them useful in the first place. The ratio of oxygen isotopes in fossils is easily altered because there is so much oxygen in water, and because there is so much water on our planet. Since the ratio of oxygen isotopes in rainwater is very negative (it is after all dominated by the isotope that is more readily evaporated, oxygen-16) the re-precipitated cement is very light too. As a result it is essential to have a suspicious nature when dealing with fossils that come from a land exposure. An oxygen isotope ratio more negative than about −4 parts per thousand is entering the area where alarm bells should begin to ring. Carbon is also to be found in meteoric waters but dissolved and in much lower concentrations. Consequently the carbon isotope ratios of fossils from land outcrops are more resistant to the big 'D'. However, in rocks with a high organic carbon content and which have a high through-flow of rainwater it is possible to alter carbon isotope ratios too. Once again care and a suspicious nature are needed. Since organic carbon is enriched in the light isotope of carbon diagenesis results in a lightening of the carbon-13/carbon-12 ratio. This parallelism between lightening oxygen and carbon isotope ratios in the presence of diagenesis means that it is possible to perform a rule-of-thumb check as the measurements are made. Most stable isotope profiles are generated from samples that are above each other in a core or rock succession. A simple correlation of one set of ratios against the other can often reveal the presence of diagenesis.

A refinement of this technique is to make multiple measurements of the same sample – say a single ammonite shell – and see if there is any correlation between oxygen and carbon isotopes. Such a correlation – if one exists – is termed a 'mixing line' and suggests strongly that the sample has been subjected to diagenetic alteration.

These though are what we might term the quick and dirty ways of checking for the presence of the dreaded big 'D'. But there are more subtle tests. The most common of these is a technique known as cathodoluminescence. A cathode ray beam is fired at a thin section through a fossil and the carbonate fabric fluoresces into different colours according to the nature of the interaction of the beam with certain chemical elements in the rock. The degree of 'quenching' (as it is rather bizarrely known) is related to the amount of manganese that is present in the rock, and manganese is commonly incorporated into calcitic shells as a replacement element during diagenesis. Consequently crimson fluorescence is an indication of the presence of diagenesis. A more subtle approach along the same lines is to measure directly the concentration of manganese as well as several other elements that are known to be diagenetically sensitive. There are several techniques available that measure elemental abundances directly; fluorescence from X-ray bombardment or from a beam of electrons can yield elemental abundances from both large samples and thin-sections and by combining data from the other diagenetically sensitive elements such as strontium, iron and magnesium it is possible to build up a comprehensive picture of the degree and nature of diagenetic alteration in a carbonate rock or fossil. In certain cases where many measurements are made on different parts of the same sample it is even possible to calculate what the original isotopic composition of the sample must have been. This is a very labour-intensive process however, one that is in marked contrast to the automated ease with which oxygen and carbon isotope analyses are now made.

It is worth mentioning that there is one class of carbonate minerals that does not require any diagenetic screening since if it is present it is by definition unaltered. This is aragonite – an unstable form of calcium carbonate whose prismatic properties give the lustre to nautiloid and certain other mollusc shells. The crystal structure of aragonite changes to that of the more stable calcium carbonate lattice very easily in the presence of the most minimal heating and through-flow of contaminating water. So if aragonite is the carbonate mineral that is measured then one knows *a priori* that diagenesis cannot have been a complicating factor.

Diagenesis – the big 'D' – can be a complicating factor not only on land, but in fossils from deep-sea environments too. The main factors here are heat and compaction. As sediment layers (originally foram ooze in most deep-sea depositional settings) accumulate, those nearest the bottom of the pile become compacted and begin to warm due to the heat of the Earth's interior.

The chalk-ooze ('chooze') region of compacting sediment is where pressure and heat are beginning to cause the carbonate to be dissolved and re-precipitated. In this environment chalk is the halfway house that is on the way to limestone.

Let's take a moment though to remember that chalk is not unique to a deep-sea environment – there is after all the Chalk of which Huxley talked – deposited in a *shallow* sea on what was the southern margin of Britain during the Cretaceous. In other words, chalk sediments can be formed in *shallow* waters if other conditions are right. There are different grades of chalk too, from a material that looks very much like ooze, through tiny fragments of replaced carbonate which adhere to the forams like icing sugar on a cake, to something that is so hard it would be difficult to mark a blackboard with it. Beyond the chalk-ooze transition, in the realm of the true deep-sea limestones, forams are only visible in thin section and are not separable from the surrounding rock. Their isotopic essence and that of the host rock have completely intermingled.

Despite the fact that the best material for isotopic studies comes from deep-sea sediments it is paradoxical that the range of measures available for identifying and controlling diagenesis is much smaller than with material from land sections. The fossils of the deep sea – for isotopic purposes, almost always forams – are simply too small to have multiple measurements performed on them. The best advice for minimising complications when working in the deep sea is to choose material from the least deeply buried sections, and look them over carefully with the optical microscope as well as the scanning electron microscope (SEM). If no evidence of the 'icing-sugar effect' is visible, most particularly if the forams have their original pearly translucency, then diagenesis is unlikely to be a problem.

The age of the deeps

Nowadays, it is hard to go a week without hearing some prediction about humankind's uncertain greenhouse future. The Second Intergovernmental Panel on Climate Change has finally agreed that mean global temperatures are rising as a result of greenhouse gases, principally CO_2, that have been pumped into the atmosphere since the Industrial Revolution. Many scientists (particularly those working in northern Europe) are especially concerned that the so-called 'North Atlantic conveyor' may shut down. The North Atlantic conveyor comprises two main currents: the Gulf Stream where surface waters travel broadly north-east from the Sargasso Sea and take warm waters back to the pole – and North Atlantic Deep Water which brings cold water south into the Atlantic at depths broadly greater than a kilometre. These current systems are important to the northern hemisphere but are only a small part of a much bigger overall scheme, the method of deep-water formation known formally as the 'thermohaline system'. The thermohaline system is where deep waters are formed by cooling of surface waters – particularly in the Norwegian Sea near Spitzbergen and the Southern Ocean near the great ice shelves of Antarctica – which increases their density and induces them to sink. These deep waters then move away from the area of their formation, sinking further into the abyssal depths, before rising to the surface in the north Pacific Ocean after a journey that has lasted about a thousand years.

This system is dependent on two things: ice at the poles which means that cooling-induced sinking is possible, and the shape of the ocean basins which imposes the route that must be taken by these deep waters.

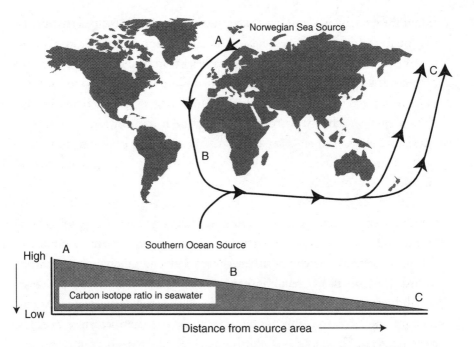

Fig. 4.3. The systematic decrease in the carbon isotope composition of deep waters as they move away from their source areas. B is more negative than A because more time has elapsed to allow the accumulation of the light isotope of carbon in deep waters. Similarly C is more negative than B.

Neither of these factors is immutable; there is good evidence that polar ice is a rare phenomenon in the geological record and of course the current configuration of the continents is merely a snapshot taken of an eternal plate tectonic dance.

Realising the temporary nature of our current climatic and topographic situation and armed with a primitive but rapidly evolving benthic oxygen isotope record which showed conclusively that deep waters had been much warmer in the geological past, two groups working in America during the eighties simultaneously asked the same question: if ice is not common in the fossil record, how and where were deep waters formed in the geological past and what effect would this have had on global climate? The question is of more than academic interest because to a very large extent the present-day pattern of deep-water circulation controls the nature of global climate. The biggest factor of all is ocean heat transport; in the present (thermohaline) situation, cold is exported via deep waters from the high latitudes to the lower latitudes and warmth is transported towards the poles in surface-water currents like the Gulf Stream. The reason why Great Britain enjoys such a balmy climate is because it is bathed in the waters of the Gulf Stream as it passes by on its way back to the Norwegian Sea. If the Gulf Stream were diverted or shut down, an immediate consequence would be that average temperatures in the UK would plummet.

In 1982 a group in Florida, led by Gary Brass, compared the benthic oxygen isotope curve with estimates of the probable area of broad, shallow seas over the last 100 million years – effectively back to the mid-Cretaceous which is the limit of the benthic oxygen isotope record. (Forams older than this have all been consumed by the processes of subduction at the ocean margins.) They found a correlation over that time period. They explained it by suggesting that deep waters in the past were not formed in the high latitudes by the processes of cooling and sinking, but rather the reverse; they were formed in the low latitudes where the preponderance of shallow-shelf seas would favour a regime where intense evaporation would lead to increased salinity of surface waters which would in turn lead to the formation of deeper waters as these denser waters themselves sank. They termed this water Warm Saline Bottom Water and the process by which it was formed and circulated became known as 'halothermal circulation'.

At the same time, a post-doctoral researcher named Tim Bralower and his mentor Hans Thierstein were realising independently that the oceans of the Cretaceous must have been radically different from the oceans of the late Cenozoic. Their paper, published in 1984, suggested that the circulation of the Cretaceous oceans was likely not only to have been powered by halothermal means but that it would also be very sluggish. This is quite different from our current ocean where the continual production of deep waters in the high latitudes maintains a rapid through-put of deep waters. The time taken today for water to circulate through the deep ocean and eventually rise again in the North Pacific is only about

1000 years. Bralower and Thierstein suggested that the average age of Cretaceous deep waters was much greater.

Was it possible to distinguish between these two models of deep-water circulation? How could the pattern of deep-water circulation in the remote geological past be found? Deep waters in the area of their formation will have a carbon-13 composition that reflects the productivity of the surface waters. If this is high, then the proportion of carbon-13 to carbon-12 (in a foram shell) will be high because much of the carbon-12 will have already been taken away by photosynthesis; conversely, if it is low then the carbon-13 to carbon-12 ratio will be low. After the water mass sinks, it is influenced only by the isotopically negative carbon that is added to it from surface waters as photosynthesising algae die and drop into deep waters. This carbon, because it is organic, reflects directly the influence of photosynthesis – and consequently has an isotopic composition of about – 25 parts per thousand (i.e. it is *very* negative). Consequently as deep waters move away from their site of formation they get older and gradually more isotopically negative with respect to carbon. In fact the present-day north Atlantic is about 2 parts per thousand more positive for carbon-13 than the north Pacific where the deep-water pathway ends.

So what does it take to measure the age of water masses? If we were only interested in the water masses now we could approach the problem by dangling a sample container over the side of a ship to the required depth and measuring the carbon isotopic composition of ocean water directly. For the geological past this is obviously not possible: we need instead to measure the record of carbon isotope compositions preserved in some form of fossil material, a form of fossil that measures only the isotopic composition of deep waters. By far the best candidates are the benthic forams who form their calcitic shells far from the surface water. Because the differences are often small, it is important to choose a species of benthic foram that is known to faithfully reflect the prevailing local proportion of carbon isotopes in sea water. Thus benthics that have been cultured in tanks by the likes of Howie Spero or Jonathan Erez and subsequently calibrated are the fossils of choice for this exercise. Fortunately benthic forams tend to evolve only slowly and modern-day calibrations are applicable over a species' entire geological range. In some cases this can be as far back as the beginning of the Cenozoic or even before, in the late Cretaceous, the era of the dinosaur and Warm Saline Bottom Water.

But having found the right fossil to measure, we find that there are still several additional complications. The most significant stems from the fact that the phrase 'deep waters' is a very loose term. After all, how deep is deep? It turns out that it is not possible to measure true bottom water, not because it doesn't exist but because there is nothing left to measure. Truly deep waters are acidic, they are below a level in the ocean known as the calcium carbonate compensation depth or CCD – technically the level in the ocean at which the rate of dissolution of dissolved carbonate balances the rate of supply from surface waters – and below it carbonate fossils will dissolve. So the CCD provides the lower limit at which it is possible to measure deep-water circulation pathways. The practical upshot of this is that it is possible to measure the movement of sub-surface waters down to a maximum of about two kilometres. In fact it turns out this is often sufficient, for it gives a hint about what is going on below.

But then there is another problem, and this stems from the depth dependency of carbon isotopes. The proportions of carbon isotopes dissolved in ocean water vary vertically as well as laterally. This vertical variation results from the interaction of two factors; first, since the rate of photosynthesis is strongly controlled by light intensity it decreases with depth. This means that the rate at which carbon-12 is removed from the water column and incorporated into organic matter decreases with depth too. Thus with increasing depth in the water column there is proportionally more of the light isotope of carbon. Secondly, isotopically light carbon that has been incorporated into living tissues in surface waters is exported from this region when algal cells die and begin the long descent into the abyssal darkness. This isotopically light carbon is returned to the ocean at a level in the water column known as the oxygen minimum zone (the precise level of this zone varies but typically it is between 0.5 and 1 km below the surface) as the organic material formed in surface waters is oxidised. The net result of these two factors is to maintain a gradient of declining carbon isotope ratios with increasing depth. The whole system can be thought of as a pump; one that removes light carbon from surface waters and pumps it to deeper levels in the water column. When this vertical variability is taken into account it becomes imperative that deep waters – or fossils that produced their shells in this deep water – be measured from equivalent levels in the water column. Effectively this means that fossil depths need to be estimated. Often this is straightforward if the site or sites were situated on a stable area of crust which has undergone very little vertical motion. But often the sites that are drilled come from areas that have not been tectonically stable – they have moved down as old crust cooled, or up as new crust formed. Under these conditions it is necessary to have some knowledge of the rate at which the crust is deforming to understand the palaeodepth of the site during the interval of interest. If this can be done, then sites of equivalent palaeodepth can be compared, and finally – finally! – patterns of deep-water circulation in the geological past can be compared.

The Post-Doc, the Co-Chief and the Box with Seven Seals. The drilling vessel JOIDES Resolution, Ontong–Java plateau, western equatorial Pacific Ocean, 01.30N, 154.50E. 20 March 1990

From the gangway the young post-doc can look down on to the red-painted decking of the core receiving area. The red non-slip surface is in places stained a patchy black from mud. Immediately to the left is the cluttered forty-foot square of the drilling floor. It is a terrifying place. Penetrating the centre of the floor the truncated oil-slicked stainless steel pipe nine inches in diameter rotates steadily, implacably, neither hurried nor indifferent, radiating a sense of quiet menace. It is attended by a dozen red-suited drillers with silver firemen's helmets. They stand around it, arms crossed, watching its rotation with a kind of wary familiarity. Others are more active, tending equipment of more enigmatic function. Slightly above the rest and to one side, another driller stands in a metal shack little bigger than a telephone kiosk, watching dials and manipulating levers with practised ease. But this place is dominated by the derrick, it rises two

Thomas Henry Huxley

Othniel Charles Marsh

PROFESSOR MARSH'S PRIMEVAL TROUPE.

Marsh's legendary fossil acquisitiveness
made *Punch* magazine

Edward Drinker Cope

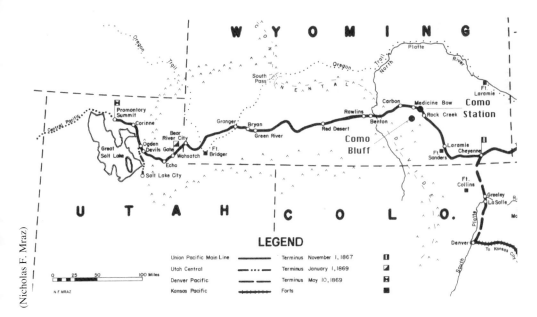

Route of the Union Pacific Railroad in 1870 showing the position of Como Station and Como Bluff

Dinosaur hunting at Como Bluff from an original watercolour by Arthur Lakes

UNION PACIFIC RAILROAD COMPANY

Agent's Office, Laramie Station [Wyoming], July 19th, 1877.

Prof. C. Marsh, Geologist.

Yale College.

Dear Sir:

I write to announce to you the discovery not far from this place, of a large number of fossils, supposed to be those of the Megatherium, although there is no one here sufficient of a geologist to state for a certainty. We have excavated one (1) partly, and know where there is several others that we have not, as yet, done any work upon. The formation in which they are found is that of the Tertiary Period.

We are desirous of disposing of what fossils we have, and also, the secret of the others. We are working men and are not able to present them as a gift, and if we can sell the secret of the fossil bed, and procure work in excavating others we would like to do so.

We have said nothing to any-one as yet.

We measured one shoulder blade and found it to measure four feet eight inches 4 ft. 8 in. in length.

One joint of the vertebrae measures two feet and one half 2½ in circumference and ten inches (10) in length.

As a proof of our sincerity and truth, we will send you a few fossils, at what they cost us in time and money in unearthing.

We would be pleased to hear from you, as you are well known as an enthusiastic geologist, and a man of means, both of which we are desirous of finding—more especially the latter.

Hoping to hear from you very soon, before the snows of winter set in,

We remain,

Very respectfully

Your Obedient Servants

[Signed] HARLOW & EDWARDS

Laramie City,

Wyoming Territory.

The original letter from Carlin and Reed. The pseudonyms Harlow and Edwards were their middle names

Gertrude Elles

Ethel M. R. Wood

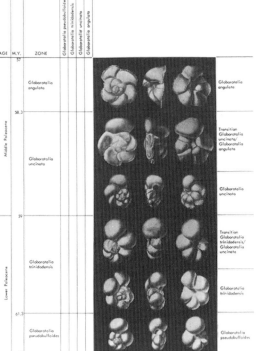

A fossil graptolite from Wenlock period rock

The evolutionary development of the Palaeocene *Morozovella* lineage

William 'Bill' Berggren

The Cavendish Laboratory, Cambridge under Rutherford. The sign refers to Rutherford's tendency to disturb the apparatus with his booming voice

Francis Aston working his mass spectrograph

Bob Savage

Jack Sepkoski

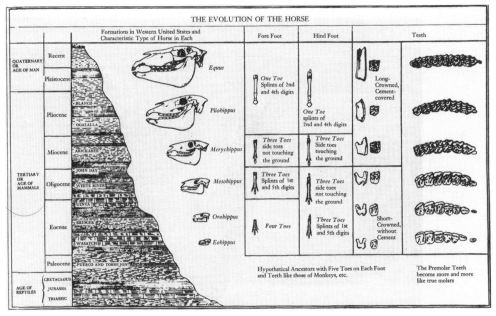

O. C. Marsh's ancestry of the horse

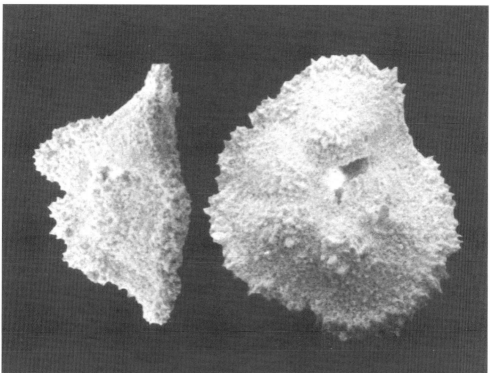

Morozovella velascoensis – a surface-dwelling form

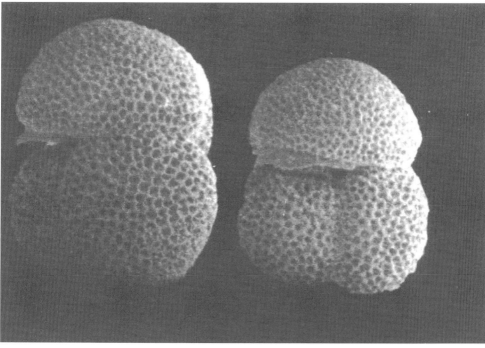

Subbotina triangularis – a deeper-dwelling form

Two species of Palaeocene planktonic foraminifera each measuring
less than half a millimetre across

Carl Woese   Linus Pauling

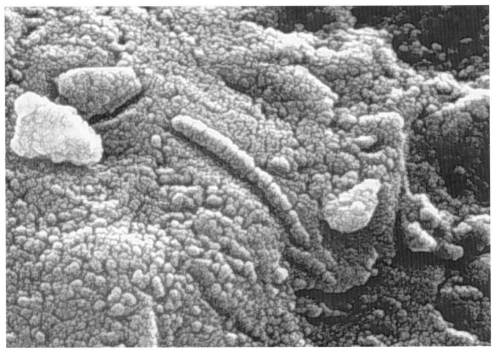

Supposed Martian bacterium

hundred feet above the drilling floor, its four massive legs rooted deep in the superstructure of the ship.

Looking up, the post-doc can see another red figure dangling from a cable a hundred feet up, hanging against the metal-framed blue of the sky like some weird circus artist. Beyond the derrick, and stretching for much of the length of the ship is the core rack, a metal box-frame containing piles upon piles of metal pipes stacked higgledy-piggledy upon each other, each waiting to be attached to the drill string. Another nine-metre length of pipe is being readied now, dangling down the length of the derrick from a thick steel cable, while the tiny figure far above swarms over and around it, checking it, prevented from the long fall to the machinery-littered metal deck beneath only by the harness underneath his backside. Now the dangling pipe is gently lowered until it touches the top of the rotating pipe, the well-head. Both ends of pipe are threaded, a male fitting matched by its female counterpart. The drillers swarm forward, galvanised suddenly into action. They grab the two pipes and heave them into perfect alignment. There is the hiss and slam of pneumatics and the drillers leap back. The new length is now rotating, and then it too starts an imperceptible descent into the drilling floor, on its way to the sea bed far beneath.

The ship has been out of Guam for over 40 days now, 40 days in which the land-lubber post-doc has had to come to terms with the new and terrifying discovery that he is slightly claustrophobic. Nothing too serious, for after all, the ship with its 2000-ton displacement is not a cupboard. No, this is all about the realisation that there is no way off this ship until it comes back to port again. The fact of the matter is that with a price tag of $2000 an hour just to keep this beast at sea, they're not going to let you off unless it's in a body bag.

The ship is at rest now, held stable only by the ceaselessly turning dynamic positioning thrusters spaced around the hull below the waterline. Beneath her dangles a length of steel pipe over a mile and a half long. It is hard to remember that only 10 miles to the west the low outline of Kapingamarangi Atoll shows up clearly in the daytime. But now it is dusk and lights are springing on all over this miniature factory. A factory ship dedicated to the advancement of marine geology. This is the JOIDES *Resolution*.

They had left Guam on the 23 January, heading south-west for the first drill site, a place that would eventually become immortalised as Hole 803, and during all of that part of the voyage, the better part of a week, he had lain on his bunk with a scopolamine patch attached under his ear and wondered just what he was doing here, 11,000 miles from home and a decent pint of warm beer. When they had arrived at Hole 803 it was time for duty, so he'd gone up to the cramped palaeo lab almost at the top of the laboratory stack, checked his microscope, and waited for the first samples to come in. Now they were over halfway through their leg and everybody's thoughts were turning towards home, the post-doc's – if the truth be known – had never really left home. But tonight he is thoughtful as he watches the drillers, mulling over a conversation that he has just had with one of the two co-chiefs. The scientific party on each leg of the Ocean Drilling Program is headed by two scientists, the co-chiefs; there are two because each takes a different shift. The whole ship runs on two twelve-hour watches, back to back, seven days a week for as long as it is at sea. And this leg is sixty-three days long. It's about as much fun as having two months of root canal surgery. This is not a conversation that the post-doc had expected to have. The co-chief in question is none other than Wolfgang Berger, known universally in the science of palaeoceanography as 'Wolf'. Wolf Berger is Mr Cenozoic – a scientist whose interest is solely in the last 65 million years of Earth's history. (Even that is a misnomer: in reality Wolf is Mr Neogene, interested only in the past 35 million years of Earth's history: the Miocene, Pliocene, Pleistocene and Holocene.) He had said as much when they were only two days out of Guam – two days, 500 nautical miles and therefore absolutely no chance of getting off the ship. 'This is a Neogene leg,' was *exactly* what he had said. And so the post-doc had been stuck aboard a ship that was effectively dedicated to drilling an interval of ocean history that held very little interest for him.

There was another cohort of scientists on board with a different agenda of course. The hard rock boys. They were interested only in basement, the deep igneous rocks that made up the true floor of the Pacific

Ocean. Those guys were on board for only two holes: the first and the last, the two slow-going holes that were going to drill all the way through the sediment column until real rock was reached. They never said as much, but you could see what they were thinking, from their whispered conversations in the hallways and the mess hall, they obviously held the same opinion as their cousins who worked the land sections. That sediments generally, fossils particularly, and the leg's Neogene objectives especially were really just gardening. And the post-doc was caught between these two camps with nothing much to do but think about the Palaeogene sediments that would have to be passed through on the way to basement and hope that something interesting would come out of it. And in truth there had been very little Palaeogene sediment that had been penetrated, so the post-doc had found himself with time on his hands. Which was why he'd started thinking about other times and other oceans: oceans that were not accessible to the Ocean Drilling Program. The oceans of deepest time itself.

It had occurred to him, with the kind of unnegotiable clarity that only forty days without a drink can bring you, that the new science of palaeoceanography was based on the oceans that Wolf and his kind found interesting – the last 35 (say 65 to be generous) million years. But what of the oceans of the deeper past? What of the oceans of the Mesozoic, the great age of the dinosaurs and atmospheric carbon dioxide levels eight times higher than at present? And what of the age before that, the Palaeozoic? It was this question, exercising his brain after twelve hours behind the microscope, that had propelled him out to the gantry above the core-receiving deck and into a conversation that he would not ordinarily have considered having with the king of Neogene oceans. Yes, the king had acknowledged, the oceans of the Palaeozoic were interesting. But the problem, argued the co-chief, was that they were so inaccessible, nothing was known about them. The stratigraphy of Palaeozoic sediments was not fine enough to compare with that of the Neogene so there was no easy way of estimating rates of processes and events. The rock coverage was inadequate, there were no great floors of ocean covered in kilometre-thick mats of ooze; instead, there were only scrubby land sections with shelf carbonates that were not representative of the deeper oceans and, of these, nothing was known. So, yes, he could see the interest in Palaeozoic oceans, but no, he could see no way of studying them. They were, he had concluded with his Neogene oceanographer's perspective, 'the box with seven seals'. And there the matter rested.

When the post-doc got home, completing a flight that would take him around the world, he would find the time to sit down and contemplate the nature of the box with seven seals. And in time he, along with several others, would decide that such a box was worth opening.

The secret ocean

This conversation took place at the beginning of the nineties when very little was known about the way that pre-KT boundary oceans had operated. The biggest problem concerned the oceans of the Palaeozoic, the time before the Permo–Triassic boundary (*see* Chapter 7), the first great era of the Phanerozoic. Its periods (tracts of time that are truly vast even by the standards of geology) are known isotopically only by handfuls of measurements and these have been made on strange and little-known organisms – mostly brachiopods – and, failing these, on the materials that now fill the gaps where once these ancient fossils were: 'cements'. Of course these cements are almost by definition diagenetic and this gives us a hint of the scale of problems posed by attempting to investigate the oceanography of the Palaeozoic. Apart from the paucity of data, the most immediate problem is that any oxygen isotope data yield results which are disconcertingly negative.

Before we investigate the reasons for the strangely negative oxygen isotopic composition of these ancient seas we need to appreciate that the study of Palaeozoic oceans is even today a woefully neglected subject; for example, a search through the Science Citation database on the Internet reveals something like only fifteen hits for the keyword 'Palaeozoic Oceans' during the period 1981 to 2000. Why is this? Why is it that no one seems interested in the oceans of the first great era of life? In a strange way, this is perhaps no surprise, it is, after all, the most recent sediments that are the most abundant and the ones with the best preservation potential. It makes sense that we should field-test new ideas in this beneficent milieu. However, ease of testing is in itself not sufficient reason. Today's funding councils are dominated by a small but influential clique of scientists who have made their reputations by studying the climatic history of the past three or so million years using the oxygen isotope technique, and it seems they can see little reason to delve further back. Their reasons are obscure. One certainly is their worry that the complicating effect of diagenesis – the Big 'D' – outweighs any potential benefits of studying the deeper past. But there is perhaps a more insidious reason – and it is worryingly self-fulfilling. The

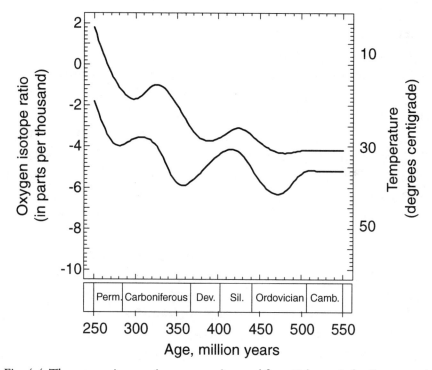

Fig. 4.4. The oxygen isotope (temperature) record from Palaeozoic fossils averaged from many different fossils and carbonate cements worldwide. The earlier part of the Palaeozoic has very much lighter values than later, probably reflecting the influence of rainfall on the shallow sedimentary basins in which these carbonates formed. The trend towards heavier values in the later Palaeozoic ('cooler' temperatures) may therefore simply reflect the gradual disappearance of these unique environments.

Pleistocene gets funded because of the apparent successes of its study, and so continues to secure funding at the expense of riskier investigations. So the study of more ancient oceans tends to be neglected.

Whether measured in fossils or cements, the oxygen isotope data from the era that precedes the Permo–Triassic boundary are very negative. In fact, as Figure 4.4 shows, the Palaeozoic can be divided into two halves just on the basis of how negative these compositions are: that portion between the Devonian and the end of the Permian (broadly the upper Palaeozoic) with compositions of around about −2 parts per thousand, and that portion between the Pre-Cambrian–Cambrian boundary and the Devonian (the lower Palaeozoic) with isotopic compositions in excess of −4 parts per thousand. Such very negative oxygen isotope ratios are quite different from those of the last 90 million years (the time between the present day and the Cenomanian–Turonian boundary) where oxygen isotopes graduate from about −4 parts per thousand to about 2 parts per thousand.

Even in the earlier Mesozoic, values similar to those of the Palaeozoic are not found. The standard explanation for such negative values as used in the Cenozoic would be to interpret them in terms of temperature. The problem is that if one does convert them into palaeotemperatures using any of the standard equations of Emiliani, Epstein or Shackleton they yield temperatures that are just too high to be credible. Oxygen isotope ratios of −6 parts per thousand equate to a temperature of 40°C: a Palaeozoic jacuzzi! The problem here as the palaeobiologist Jim Valentine has pointed out, is that organisms simply cannot survive temperatures this high. Since animals unarguably inhabited the seas of the lower Palaeozoic, the reason for these very negative values is unlikely to be related to temperature.

With our understanding of oxygen isotopes, two other explanations present themselves immediately: first, the big 'D' – all measurements that stem from the ancient rocks of the Palaeozoic have been altered – there is nothing left to give us a true estimate of the temperature of these ancient oceans. To answer this we need to fall back on the bugbear of the case lawyer: circumstantial evidence. The argument is this – wherever we have been in the world to measure the oxygen isotopic composition of Palaeozoic rocks this range of values is found – generally speaking more negative than –4 parts per thousand. It beggars belief that *every* Palaeozoic limestone that has *ever been measured* should be altered, surely? And that leaves us with the final category of answers – that the original oxygen isotopic composition of the Palaeozoic oceans was more negative than that of either the Mesozoic or the Cenozoic. And this of course is the ultimate facer – for what can have caused the oceans of the Palaeozoic to be so much more negative to start with than the more recent oceans that we are familiar with?

A possible answer lies in the palaeodepth from which these measurements are derived. All Palaeozoic measurements are taken from carbonates that were deposited in relatively shallow shelf seas – anything from a few tens of metres to perhaps a hundred metres deep, depths which are quite different from the depths at which the deep-sea limestones that we are familiar with in the Cenozoic were deposited in. For example, the wonderful limestones of the Bottaccione Gorge (*see* Chapter 6) are all deep-sea limestones, deposited in perhaps a kilometre or more of genuine deep water, while the limestones of the Welsh borderland were deposited in probably less than 200 metres of water. This is the essential difference, and complication, of comparing Palaeozoic and post-Palaeozoic sea water using the oxygen isotope technique. We are not comparing like with like.

Several authors have noted this in different ways and yet there is a similarity in their solutions. The first was Bruce Railsback who, as a graduate student, was attracted to the study of these most ancient seas through the interests of his graduate supervisor, Tom Anderson of the University of Illinois at Urbana. Railsback studied the ancient limestones of the Trenton group of New Jersey and environs. This is a locality that has much in common with the lower Palaeozoic limestones of Britain because both sequences were deposited on either side of the same ancient ocean – the Iapetus – and in the Ordovician this was an ocean that was beginning to close rapidly. Railsback started from the premise that there was only a limited range of salinities and temperatures that were possible in any ocean, modern or ancient, and then he went on to constrain the combination of both that was needed to produce the negative values of the Ordovician. He concluded that to produce values as light as –6 parts per thousand the water must have been both very warm and significantly saline. He concluded in effect that the dominant type of deep water in the Ordovician ocean was of the Warm Saline Deep Water type. This provided the perfect explanation for the very light values of lower Palaeozoic limestones, for they were all deposited in shallow waters, and these shallow waters would be artificially enriched in the light isotope of oxygen: oxygen-16. The formation of deep waters on the continental shelves of these ancient seas by evaporation led to the net export of oxygen-18 to deep waters, beyond our ability to measure.

This explanation spawned two modifications almost simultaneously. Two groups, one in Canada at the University of Saskatchewan led by Chris Holmden, the other in Great Britain led by Peter Allison of Imperial College, made the observation that Palaeozoic limestones are not merely deposited in shallow waters, they are also laid down in a particular type of shallow ocean – epeiric seas. Epeiric seas are a type of water body that is not to be found as part of the oceans of the present day; they are the result of the relatively high sea levels of the Palaeozoic that flooded large portions of the continents. For example, much of eastern North America is a very large, very stable tectonic block known as a craton. During the Ordovician, this region was flooded by a part of the Iapetus Ocean known as the Mohawk Seaway, which is why it is now so richly fossiliferous and consequently the focus of so many studies into Palaeozoic palaeoceanography. But the essential feature of epeiric seas is that they are very shallow – perhaps only a few tens of metres deep. This is why there are so many wonderful limestone deposits from Palaeozoic epeiric seas (the limestones of the Welsh borderland are a good example), for such limestones are formed by coral reefs – modern-day analogues of the reefs and atolls of the Bahamas and the South Pacific – that can only grow in relatively warm, sunlit waters.

Flooded cratons can be thought of as a series of dew pools but on a much larger and more permanent scale – a series of shallow, pond-like seas spread across vast tracts of continental crust. The Holmden and Allison groups posited that such ponds would be easily susceptible to oxygen isotopic alteration by rainwater run-off from the surrounding basin margins. To test this idea, the Holmden group used carbon isotope distributions in the Mohawk and Taconic Seaways to see how well waters circulated. They concluded that these basins were not only sluggish but had very little connection with either adjoining basins or the deep waters of the nearby Iapetus Ocean. The Allison group took a broader view, using a mathematical and modelling approach, and concluded much the same thing.

And so the situation rests; the reasons for the very negative oxygen isotope values of the Palaeozoic oceans are not because of either extreme temperatures or even pervasive diagenesis. The reason that the oxygen isotope values of Palaeozoic oceans are so light is much more to do with the fact that nearly all our measurements of this ancient time – in many ways as remote and fabulous as a different planet – come from sediments that are not analogous to the sediments we routinely use to measure the temperature of more recent oceans. We are simply not comparing like with like. Perhaps future work based on the techniques of measuring the oxygen isotope composition of phosphates in Palaeozoic invertebrates or fish remains will reveal more, but until then the oceans of the Palaeozoic remain a book with seven seals.

The Savage Hand of Evolution

A term starts in Bristol

The sun is slanting in from the south-west. Off across Kingswood and the city slumped in its depression I can just see the Mendips hulking like guardians in front of the distant brooding presence of Exmoor. Immediately though, down the hill that slopes so remarkably into Bristol, the buildings of the chemistry department descend in their untidy fashion. Their grip on the hillside seems somehow worried, as though they are not sure about being ranked up the flank of St Michael's Hill in defiance of gravity in this way. That's how Bristol has always seemed to me. Crammed together, untidy, yet supremely confident.

In front of me, as I stand in the doorway, is the palaeo lab. I'm not sure what I expected. I've been used to zoology, botany and physiology labs for the past year. But the palaeo lab is something else. A perfectly blank table stretches the length of the room. There is a blackboard at the far end and a token sink stands in the corner between it and the big picture window. On the other side, on the long wall, a bank of drawers filled with fossils stretches from floor to ceiling. The drawers smell of polished wood. Where are the conventional lab benches with their spigots for water, oxygen, gas? Where is the untidy clutter of experiments in progress, the gentle presence of the senior lab tech in zoology – a wonderful old chap called

Albert – as he bustles with his heavy spectacles between experiments and this year's undergraduates? Where is the smell of formaldehyde and the other nameless odours of zoology labs everywhere? I'm beginning to realise that the palaeo lab isn't like that and that palaeontology isn't like other sciences either. Both seem simultaneously enigmatic and yet serene, as though they are about some other business altogether.

And of course they are.

The palaeo lab is in an entirely different department. This is geology, and I'm only here because of some fast talking on my part at the end of the first-year course. Last year they showed us most of David Attenborough's *Life on Earth* as part of our zoology practical class, and somewhere amongst it and the pigeon hearts and the liver flukes I've realised what biology is all about. It's been my own personal revelation. I've even had arguments with my neurophysiology tutor – a man whose zoological interests lie exclusively in the details of communication between nerve cells – about it. It's an early example of what I will soon be told is my bad attitude (it's also a habit I've never chosen to break; the only stupid question is the one never asked). I'm hooked on the enigmatic foundation of all biology, the bit that never seems to be talked about by biologists, but which is what they tacitly rely on to make their subject work. I've become an evolutionist. And like zealots everywhere I've been a bit of a pain in the backside to get what I want. So suddenly I'm in the University of Bristol's Department of Geology, and I'm a bit unnerved because the palaeontology laboratory doesn't look like any laboratory I've ever seen before.

Palaeontology is the subject that I need to spend time with in order to understand evolution. I'm trying to find the Unified Field Theory of Biology (in those days I was too young to be awed by the arrogance of a thought that big) and I know that it will only be found where there are the remains of evolution to view directly – in the fossil record. Which is why I've decided to rejig the focus of my studies, and have managed to finagle myself into Bob Savage's second-year palaeontology course. Yet as I stand there gazing into a new laboratory and a new subject area I've no idea that only a few years before this place was central to one of the most controversial debates in modern evolutionary palaeontology.

The debate was over the rate of evolution as seen in the fossil record. In the jargon of palaeontology this came to be known as the 'punctuated equilibrium' versus 'phyletic gradualism' debate. The latter was the accepted wisdom, the former the big new idea promulgated by the young Americans Niles Eldredge and Steve Gould. It was never mentioned overtly – that Peter Williamson and Bob Savage had produced just about the earliest proof of Eldredge and Gould's seminal hypothesis. You just had to sort of soak it up, be in the know, be part of the Bristol Mafia. Looking back on it now I can see clearly how Bob ran his combined geology/zoology course very much like an exclusive London club. In the year I partook of his tutelage there were precisely two other undergraduates actually enrolled in the course. There were also only two graduate students; one specialising in the dentition of rodents, and the other studying the strange Triassic bone-beds of Aust cliff, which lie just underneath the English side of the old Severn Bridge.

To this day I still don't know why Bob Savage took me on.

Microevolution vs macroevolution

In 1972 a obscure paper was published in an obscurer book: Tom Schopf's *Models in Paleobiology* (what do you mean you've never heard of it!) that was to change the world of evolutionary palaeontology. For some reason Niles Eldredge and Stephen Jay Gould decided to publish their paper 'Punctuated Equilibria: An alternative to Phyletic Gradualism' in this book. Eldredge was a junior professor at the American Museum of Natural History and Stephen Jay Gould was not yet, well, Stephen Jay Gould. In those days Gould was 'merely' regarded as the brightest young palaeontologist in the world, a man who had made his reputation at Columbia University in New York studying the mathematics that underlies the construction of fossil vertebrates (his review paper on this subject, 'Allometry and size in ontogeny and phylogeny' remains a classic to this day). Steve Gould was never a man to be awed by the occasional big thought and so in the seventies he turned his attention to the rate of evolution as it was measurable in the fossil record, and that was when he, in conjunction with Eldredge, dreamed up the concept of punctuated equilibrium. To understand the significance of the punctuated equilibrium theory we first need to understand the essential problem of distinguishing what is called macroevolution from micro-evolution, and to do this we need to revisit the history of evolutionary biology.

Evolutionary biologists and palaeontologists have a special name for the external appearance of an organism be it animal, plant, bacterium or indeed anything else. They call it the *phenotype*. The phenotype also refers to any and all organs and the physiological processes which go on within them to sustain life. In short, phenotype refers to the physical manifestation of a living organism. The information that controls the phenotype, the genes, is known as the genotype. The architect of eternity who discovered the rules governing the transmission of genetic information was the monk Gregor Mendel. Mendel, working in the Augustinian monastery at Brunn, Moravia in the 1840s and 1850s, showed that there were two main types of gene, a dominant gene and a recessive gene, and that the way they were combined in offspring determined their expression in the phenotype. If either parent passes on a dominant gene to its offspring then that offspring will always show the phenotypic character coded for by the dominant gene. Only if both parents donate recessive genes to their progeny would that offspring show this recessive phenotypic character. Mendel had no notion of how this information was either encoded or transmitted. That mechanism was not discovered until almost a hundred years later when James Watson and Francis Crick made their epochal discovery of the structure of the DNA molecule in Cambridge. Indeed, although Mendel was aware of Darwin's mechanism of evolution by natural selection, he chose to publish his own work in an obscure local journal that meant its significance was buried for several decades. The connection between his laws of heredity and Darwin's theory of evolution was not made until the turn of the twentieth century when Mendel's laws were rediscovered by, among others, the great Dutch naturalist Hugo de Vries.

De Vries originated the theory of 'mutationism' that asserted that new species arise by major changes in the genetic structure of an animal or plant. This was quite at odds with Darwin's original notion that the minor differences between parents and offspring that were well known from ordinary animal and plant husbandry were the basic fuel of the evolutionary process. The mutationists (or Mendelians, as they were also known) were opposed by other thinkers like the great Victorian scientist Karl Pearson. Pearson was basically a mathematician but was also a polymath. He ran a law practice in the early 1880s and then in 1884 was appointed Professor of Applied Mathematics at University College London where he became Professor of Geometry, and head of the department of applied mathematics, as well as Galton Professor of Eugenics! From about 1893 onwards, Pearson became interested in using statistics to understand the problems of evolution and heredity. Pearson and his associates defended Darwin's view that evolutionary change was effected by the normal processes of genetic change; that is, small variations that accumulated across the generations until large-scale evolutionary differences had been manufactured. In this respect Pearson can be seen as the originator of the 'microevolution' school of thought: tiny genetic differences given time can power macroevolution, the creation of new species. Members of Pearson's school came to be known as the 'biometricians'.

Although in Mendel's day the experimental organism of choice was the simple garden pea that changed in the early years of the twentieth century through the experiments of the American Thomas Hunt Morgan. Morgan, born in 1866 the nephew of a confederate general, showed an early aptitude for natural history and received a bachelor of science degree from the University of Kentucky in 1886. After entering Johns Hopkins University in Baltimore for graduate work he was awarded his PhD in 1890. He accepted a post at Bryn Mawr College in 1891 where he stayed for over a decade, eventually marrying one of his graduate students, the talented cell specialist Lillian Sampson in 1904. That same year he was offered and accepted a professorship of experimental zoology at Columbia University in New York City. It was at Columbia that he became interested in using the fruit fly *Drosophila melanogaster* as an experimental vehicle for research on the genetic basis of heredity. At that time the chromosome theory (which states that the chromosomes are the carriers of genes) was just beginning to gain currency in the scientific world. However, Morgan himself was very sceptical about the validity of the chromosome theory. The advantage of *Drosophila* to these early pioneers was that the chromosomes of their salivary glands are huge, and were therefore easy to study with the light microscopes available at the turn of the twentieth century. Another advantage of *Drosophila* is its amazing fecundity, with new generations being produced every two weeks!

Morgan began his experiments on *Drosophila* in 1908 and within a year had noted a phenotypic variation known as 'white-eye' in a single male within one of his breeding bottles. Morgan bred this male with ordinary red-eyed females and noted that all of the offspring were red-eyed. Matings between the members of *this* generation produced a second generation, some of whom had white eyes. The white-eyed variants were all males. This led Morgan to develop the hypothesis of sex-linkage, which is that certain inheritable traits were carried by one chromosome that was not shared between both sexes. He suggested that this was part of the X-chromosome of the female fly.

The work of Thomas Hunt Morgan was to be the catalyst around which the so-called 'synthetic theory of evolution' would eventually coalesce. Almost immediately after he had started his work on *Drosophila*, graduate students began to join him. The greatest of these were Calvin Blackman Bridges and Hermann Joseph Muller. Bridges entered Columbia University in 1909 and only a year later, inspired by Morgan's work, had obtained a position as laboratory assistant to the great man. Muller, who had been studying at Columbia since 1907 becoming progressively more fascinated by the discoveries that were beginning to pour from Morgan's lab, similarly joined the team in 1912 as laboratory assistant. Over the next decade Morgan, Muller and Bridges produced a series of classic papers that shaped twentieth-century genetics. Muller's work, for example, proved the linear linkage of genes along chromosomes and earned him his PhD in 1916. Bridges showed that the phenotypic variations in *Drosophila* populations could be traced to changes in the structure of an animal's chromosomes. These two, together with Morgan and his other graduate students like Alfred Henry Sturtevant, established the chromosome theory of heredity. Morgan summarised this work in two classic books: *The*

Mechanism of Mendelian Heredity in 1915 and *Sex in Relation to Chromosomes and Genes* in 1925. Hermann Muller went on to become a professor at the University of Texas in Austin where he remained until 1932. It was during this period that he proved that X-rays could induce changes in the order of genes along chromosomes; that is, that they could cause mutations. This work eventually won him the Nobel prize for Medicine in 1946 (not to mention spawning a whole pulp sub-culture about mutants and radiation immortalised in the 'B' movies of the forties and fifties).

But through a combination of overwork and personal problems Muller suffered a nervous breakdown in 1932. To help him rehabilitate he spent a year in the Max Plank Institute in Berlin and then moved on to Moscow at the invitation of Nikolai Vavilov, head of the All-Union Academy of Agricultural Sciences. Muller was a socialist who looked on the Soviet Union as a progressive society which could well be a model for the future of mankind. But Joseph Stalin had come to power in 1929 and Soviet science, particularly genetics, under the twisted guidance of Trofim Denisovich Lysenko was just about to take a strange and unhealthy path. Lysenko had graduated from Uman school of horticulture in the Ukraine in 1921 and since then had been posted at various Soviet experimental agricultural stations. Exposed to the emerging ground-truths of Soviet Communism and fuelled by an unbridled ambition, it was not long before Lysenko betrayed both himself and his science. Realising that the way to get ahead in Joe Stalin's Brave New World was to say the right thing, he publicly renounced the new genetics coming out of the West from places like Morgan's lab in America in favour of the doctrines of a Russian horticulturist named Ivan Vladimirovich Michurin, who had attempted to resurrect the nineteenth-century theory of the inheritance of acquired characteristics (so-called Neo-Lamarckism). The inheritance of acquired characteristics was favoured by the new Soviet regime because it gave a spurious scientific credibility to the regime's political underpinning of dialectical materialism. Human societies could improve themselves and pass on these improvements to succeeding generations: socialism and communism were the natural end points of human political development.

By the time Muller had arrived in the Soviet Union, Lysenko had been appointed a senior specialist in the department of physiology at the All-Union Institute of Selection and Genetics in the Ukraine. In the 1930s the Soviet Union experienced an agricultural crisis of stupendous proportions. There was not enough food to feed the population and the regime's credibility was in jeopardy. The great Soviet social experiment showed signs of collapse. It was Lysenko's golden opportunity. Michurin, despite being an uneducated man, had successfully hybridised more than 300 new types of fruit trees and berries which he had used as proof of his strange theories. Lysenko promised that under his leadership crop yields would show spectacular improvements and the hungry mouths of the proletariat would be filled. However, even under the growing oppression of Stalinism, Lysenko's extravagant claims were challenged. Chief among his opponents was Vavilov who, having studied under William Bateson (one of the founders of modern genetics) at Cambridge, knew full well just how far removed Lysenko was from mainstream genetics. And seeing his dream of a socialist utopia being smashed by this false prophet, Vavilov's guest Muller joined him in discrediting the new genetic doctrine.

Muller was eventually forced to leave the Soviet Union in 1937. Lysenko's repeated denunciations of Vavilov at scientific meetings in the late thirties in the end destroyed his rival's reputation. Vavilov was arrested in 1940 and imprisoned in a concentration camp. The same year that Vavilov was arrested Lysenko was appointed Director of the Institute of Genetics of the Academy of Sciences of the USSR where he stayed for the next 25 years. He also took over Vavilov's old job as director of the All-Union V. I. Lenin Academy of Agricultural Sciences. Vavilov died in prison early in 1943. Lysenko's appointment to Vavilov's old job marked the start of some of the worst scientific witch-hunts in history. By 1948, research into classical genetics had been outlawed and those who would not capitulate imprisoned and often executed. For three decades Lysenko forced Soviet agriculture along a dead-end trail with no scientific credibility whatsoever. It would not be until 1964 that Lysenko's doctrines would be discredited and he and his followers stripped of their power. Only now is genetics recovering in the one-sixth of the world that used to be the eastern bloc. Lysenko may be discredited but his legacy lives on in one of the finest pieces of Western scientific fiction ever written. In 1951, John Wyndham wrote *The Day of the Triffids*. And the terrifying man-eating vegetables about which the book revolves were created, you guessed it, by Lysenkoist genetics. Food, if ever there was, for thought . . .

The Trinity

In 1928, four years before Muller had his nervous breakdown and set out on the path that was to involve him in the strange story of Trofim Lysenko, his old mentor Thomas Hunt Morgan was invited to set up a Division of Biology at Caltech (the California Institute of Technology) in Pasadena. He took with him both Calvin Bridges and a new student who the year before had come to him from, strangely enough, the Ukraine: Theodosius Dobzhansky. Despite hailing from the same region of the Soviet Union as Trofim Lysenko, the impact that Dobzhansky was to have on evolution and the new science of fossils could not have been more different. Theodosius Dobzhansky is a true architect of eternity. For infinitesimal changes in a genetics lab ('microevolution') is very far from the same thing as understanding the sequence of fragmented horse-skeleton remains strewn across the Wyoming badlands, or tiny changes in graptolites in rock sequences in the Welsh borderland. These are examples of 'macroevolution', and before Dobzhansky, to connect the two required nothing less than a leap of faith. Dobzhansky, building on the groundbreaking work of Morgan and his students, was one of three scientists who would connect the two and establish the so-called 'Synthetic Theory of Evolution'. Under the 'Trinity' the mechanism by which information was passed between generations and populations would become sufficiently well understood to attempt to join it with the pattern of evolution as seen in the fossil record.

Along with the experimental work done by Morgan and his school, others had been considering genetics from the more theoretical standpoint of the statistics pioneered by the polymath Karl Pearson. British evolutionists R. A. Fischer and J. B. S. Haldane teamed up with others, such as Sewell Wright in the United States, to show how changes in the genetic code could effect phenotypic differences in succeeding generations and go on to spread through populations. However, until Dobzhansky, their work was largely ignored because it was written in a mathematical language that contemporary biologists had simply not been trained to comprehend. The breakthrough – the joining of the theoretical and practical schools of genetics – came in 1937 when Dobzhansky wrote his masterpiece, *Genetics and the Origin of Species*. It established the genetic basis of the synthetic theory of evolution. But Dobzhansky could not work alone. To forge a true synthesis of evolutionary theory he needed to connect his new unified genetics to both biology and palaeontology. And, as is so often the case in the history of science, two people were waiting in the wings to assume just this role: the biologist Ernst Mayr and the palaeontologist George Gaylord Simpson. The problem that this trinity of evolutionary intellects faced was fundamentally one of timescale. They had to link the microevolutionary genetic mechanisms of *Drosophila* to the macroevolutionary changes of the fossil record.

German-born Ernst Mayr received his PhD from the University of Berlin in 1926. Profoundly interested in ornithology from an early age, Mayr, appointed to the university staff at Berlin after completing his doctoral work, led the first of three expeditions to New Guinea and the Solomon Islands in the late 1920s. There he became impressed with the geographic variability of bird species and he started thinking about how species might be formed in the wild. He suggested that species were formed when small 'founder' populations became split off from the main population. Over the next two decades he developed this theory into the 'allopatric' model of speciation which states that it is easier for small isolated populations to form new species than the larger, main population because there is less 'genetic inertia' to overcome. So Mayr's contribution to the synthetic theory of evolution was to show how genetic changes could flow through populations (if they were small enough) with sufficient speed to cause new species to form.

In 1932 Mayr was appointed curator of birds in the American Museum of Natural History in New York. It was there that he met George Gaylord Simpson who, having received his PhD from Yale in 1926, had in

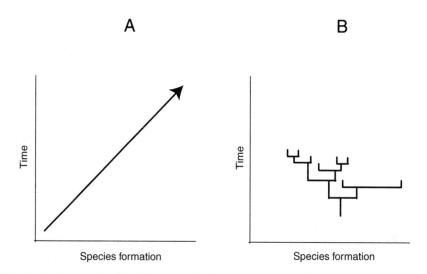

Fig. 5.1. George Gaylord Simpson's ideas about how species evolve. A. The 'phyletic' mode, whereby one species gradually changes through time into another. He regarded this as the most common form of species formation. B. The 'speciation' mode. Simpson thought this a much rarer form of evolution but we now know that it is probably far more common than the phyletic mode. Speciation formed the basis of the 'punctuated equilibrium' idea of Niles Eldredge and Stephen Jay Gould.

1927 himself been appointed to the staff of the palaeontology department. Simpson became interested in how the ideas of Dobzhansky and Mayr could be applied to the fossil record and so account for the ways that species were formed over geological time. His ideas led him to propose three modes of evolution: 1) the 'phyletic mode' whereby species evolve infinitesimally into one another through time (*see* Fig 5.1); 2) the speciation mode (much rarer, or so he thought, accounting for perhaps only 10 per cent of new species formed in the fossil record), where a lineage splits to form two species; and 3) the saltational mode which accounts for the sudden appearance of higher taxonomic groupings – genera, orders, classes etc. Together the Trinity enshrined the so-called 'Neo-Darwinian paradigm' (paradigm is simply a fancy name for 'model') and placed it at the heart of the synthetic theory. This Neo-Darwinian paradigm states that: 1) more individuals are born than can survive; 2) these individuals are neither genetically nor phenotypically identical, the variation being fuelled by the usual mix 'n' match of genetic recombination which occurs when sex-cells are formed; and 3) those individuals best suited (or 'fitted') to the environment will survive, while those less 'fit' will fail to survive. Simpson acknowledged that his saltational mode might require some mechanism *other* than the neo-Darwinian paradigm – but was insistent that phyletic and speciation modes could be adequately explained by extrapolating from the mechanisms of genetics as understood at the level of the individual (i.e. *Drosophila*) and the population. In other words, Simpson, even while promulgating the synthesis, acknowledged that the most important part of it – the evolution of new body plans and higher taxa – was still so imperfectly understood that it still required nothing less than a leap of faith.

Simpson, Mayr and Dobzhansky wrote three books between 1930 and 1950 which effectively defined the synthetic theory of evolution: as well as Dobzhansky's *Genetics and the Origin of Species* in 1937, Mayr wrote *Systematics and the Origin of Species* in 1942 and G. G. (as he liked to be known) wrote *Tempo and Mode of Evolution* in 1944. These three books explained the mechanism of macroevolution as the mechanism of micro-evolution but on a larger stage.

Implicit in the synthetic theory is the notion of gradualism: that the absolute requirement for evolution to work is trackless quantities of geological time. This notion of gradualism goes back as far as Darwin himself. Darwin was preoccupied with the notion of gradualism, the idea that change – particularly in human societies – could only come about slowly. The notable exception to this idea was Thomas Henry Huxley, founder of the new science of fossils, who chastised Darwin even before the *Origin's* publication for his insistence on gradualism in the evolution of new species. The idea of gradualism was exacerbated by other practitioners of the geosciences because they were so influenced by Charles Lyell's *Principles of Geology* which had recently and explicitly enunciated the Principle of Uniformitarianism (events in the past are not unique but are the same as those occurring in the present day; they were simply more spectacular because of the vast quantities of time over which they had been able to operate). And when Dobzhansky, Mayr and Simpson came to create the synthetic theory, their world view had not yet changed sufficiently for them to see evolution in any way other than as a gradual process that ground slowly – but exceeding fine – over vast amounts of geological time. In hindsight this seems strange, for while Dobzhansky had been working with Morgan he had shown that changes in the genetic structure of *Drosophila* populations could happen very quickly, over just a few generations. It may be that he sacrificed the idea that evolution on geological timescales could occur rapidly in order to ensure support for the synthetic theory.

'Fitness', by the way, is the standard shorthand for organisms that are most in tune with their environment, the textbook example being the peppered moth *Biston betularia* whose mutation to darker pigmentation allowed it to survive in Britain's industrial heartlands while its lighter coloured ancestor hit the evolutionary bin. The synthetic theory states that this differential in average fitness drives a species' evolutionary success and means that only those best fitted to the requirements of their environmental niche will survive. Subsequent generations are driven inexorably towards greater and greater fitness and so evolution of progressively more refined organisms occurs. Evolutionary palaeontologists have a special name for successful body designs that are subsequently worked on by the refining fire of natural selection; they are called *bauplans*.

The adaptive landscape

New York City in the forties and early fifties was the spiritual home of the synthesis. Dobzhansky had returned to Columbia from Pasadena in 1940 as Professor of Zoology, and Mayr and Simpson were well installed at the American Museum of Natural History. So the Gospel of the Trinity filtering down out of the Big Apple was that macroevolution and microevolution had now been properly connected and that evolution could be explained at almost every level by the processes of known genetics. The watchword of the synthesis was 'gradual'. Why was this? Part of the reason was the pressure that the Trinity must have felt to gain acceptance for the synthetic theory of evolution. They were also, particularly Simpson, aware that the man himself – Darwin – had stated explicitly that his theory lived or died by the sword of palaeontology and they must have felt the pressure to produce results from the fossil record.

According to the Trinity, particularly Simpson, the 'phyletic' mode of evolution was the dominant one (speciation and saltation being much rarer) and therefore transitional forms should be found in the fossil record. But, like Darwin before them, they were having to acknowledge that for the most part the fossil record was unable to produce the goods. There were exceptions of course and these were widely toted as proof of evolution. One of the most famous was the ancestry of the horse as documented by Marsh. There were other examples as well. The spectacular *Archaeopteryx* fossil from the Jurassic seemed perfectly placed as the halfway mark between a bird-like ancestor and true birds; the small worm-like creature *Peripatus* found extant in the tropical rainforests of south-east Asia had some of the characteristics of annelids (earthworms and their ilk) and arthropods (the group that includes crustacea and insects). But by and large, transitional forms and missing links were hard to come by, especially in the fossil record.

Why should we expect transitional forms? For that matter, what had persuaded Simpson, Dobzhansky and Mayr that the processes of microevolution could account for macroevolution? The answer lies in a picture, or more accurately a pictorial metaphor developed by one of the unsung heroes of the synthetic theory: Sewell Wright. He was a geneticist interested in the way in which gene flow in populations might lead to phenotypic variation and allow evolution to function. He saw animals and plant populations of a species as groups of genes in a vast landscape of hills and valleys. The populations are separated because they do not share identical genetic compositions. The hills and valleys are a metaphor for fitness – the hills are an organism's 'optimal' tuning to the environment, the valleys are the regions where tuning is sub-optimal, the deeper the valley the more sub-optimal the tuning. In other words, the further a population lies from a peak the less well-adapted it is. To take an example from the avian world, a peak might represent the best fitness for long-distance flight. Under Wright's scheme an ancestral population of genes will progressively modify as mutation and recombination throw up variants that are progressively better suited to this lifestyle. (This pattern of variation plus movement of a population of genes towards increased fitness is called 'selective pressure'). Ultimately, you arrive at an albatross, with a seven-foot wingspan ideally suited to fly vast distances above the southern oceans (until, that is, it meets the ancient mariner, or modern-day fishing methods).

It is clear then that the peaks are really a convenient shorthand for the degree of success in filling an ecological niche, i.e. how well adapted an organism is. Wright's pictorial metaphor was originally generated as a vehicle for highlighting the problem of how genetic populations go about moving from the valleys to the peaks. Do only a couple of semi-formed flying feet enable a tree frog to jump effectively between trees? How useful is half an eye? Successful niche occupancy dictates that you either can do it or you can't; there is no halfway house (or as Admiral Jonas Ingram once famously remarked, 'We got no place in this outfit for good losers – we want tough hombres who will go in there and win!'). But it was seized on by the Trinity as *the* visual tool that could explain the whole of the synthetic theory of evolution and would enable them to sell the idea. Simpson, in *Tempo and Mode* took Wright's image of within-species gene flow and scaled it up just as he had done with the notion of microevolution and used it to explain the differentiation of life at higher taxonomic levels (genera, orders, classes etc). Simpson, by scaling up the concept, also reinforced the metaphor of the landscape, for the location of the peaks and valleys was always changing as the environment evolved and the competition for those peaks from other evolving organisms changed. This must have appealed to his palaeontologist's soul, for the adaptive landscape then became a metaphor for Earth's own landscape with seas and lakes expanding and receding, with mountains being thrust up, and so on all across enormous quantities of geological time. Populations or organisms were conceptualised as literally fighting for success by moving across Earth's constantly changing surface gradually changing into new species as the goal-posts imperceptibly but eternally moved. By 1951, Dobzhansky was using Wright's image to explain the evolution and relationships of all of life itself. He visualised the entire menagerie of Earth's population of animals, plants, bacteria etc. as occupying hills and mountain ranges on Wright's adaptive landscape. And these mountain ranges explicitly reflected the scheme of classification of all life that was then in vogue.

The Trinity favoured the so-called 'phylogenetic scheme' whereby all of life was pigeonholed (i.e. related) according to its supposed evolutionary history. Thus the chordates – those higher animals with a stiff rod (a notochord) in their backs at some point in their life cycle – would occupy that portion of the adaptive landscape that accommodated all those animals based on the notochord *bauplan*. Within this, mountain sub-ranges would accommodate the vertebrates: the fish, amphibians, reptiles and mammals. These sub-ranges in their turn would accommodate all the plethora of types within these taxa, for example, within the mammals, three additional sub-ranges would accommodate the egg-laying mammals such as the platypus, the pouched mammals or marsupials such as the wombat and kangaroo, and the group of mammals of which we are a member where foetal growth occurs within the body. The adaptive landscape then is the ultimate pictorial metaphor that describes the synthetic theory of evolution: a vast and endlessly changing spectrum of ecologies ceaselessly tracked by evolving communities. The endless, gradual ballet of life.

'Punk eek'. Boston, MA, USA, 42.20N, 71.05W. Sometime in the early 1970s

Outside the hospital the sound of horns and sirens has become blurred into an indefinable wall of sound. But Steve Gould, his head wrapped in bandages, doesn't mind. There's not much that can dent his Kevlar self-confidence at the best of times and today is definitely the best of times. He has just been told that the squash-ball injury he sustained to his eye will leave no permanent damage and, in addition, an idea that he's been kicking around with Niles Eldredge has crystallised and confirmed what he's been suspecting for some time. There's a hole a mile wide at the centre of current evolutionary thinking.

The theory of evolution – even in its synthetic guise – is fundamentally flawed and the problem goes as far back as Darwin himself. It's not so much the science as Victorian mentality that has led evolutionary biology astray. Darwin believed in the concept of gradualism in society and this spilled over into his palaeontology. Change, such as improvement in working conditions and education for the masses was desirable but only as a gradual introduction. Rapid, uncontrollable changes were destructive and to be discouraged at all costs. To Gould it was all suddenly very clear: because the Victorian scientists who originated and developed the synthesis were social gradualists, they assumed that evolution would occur gradually, too, a classic case of prevailing culture colouring scientific thought. But why does evolution have to proceed slowly? What's to stop it occurring rapidly by some as yet unknown mechanism? Most particularly, what does the fossil record show? The pioneers of the synthesis (even Darwin himself) had stated repeatedly that the fossil record was too gappy to show more than a few isolated examples of transitional forms. But that stemmed more from Darwin's and the Trinity's preconceptions of how evolution occurred – by gradual transformation of ancestral types into descendent forms – than anything else. But what if evolution isn't like that? What happens if evolution proceeds as bursts of species formation in semi- or completely isolated populations, where radical change can spread rapidly through a small population, in other words, where the inertia of the prevailing genetic make-up is drastically reduced? And you had to face it: the number of examples of phyletic transformations found in the fossil record was extremely small. Close observation of the data showed that in most lineages there were long periods when nothing much happened and the appearance of a new form happened almost overnight (at least, in geological terms) . . .

That was what Steve Gould was thinking about that day in the early 1970s. He and Niles Eldredge had developed the idea of punctuated equilibrium – or 'punk eek' as it would come to be known in the popular vernacular – and had just published it. Now Gould was working out the implications. When he and Eldredge published the idea in *Models in*

Palaeobiology, they formalised the notion that a novel genetic change could spread rapidly through a population in terms of the allopatric speciation model of Mayr – literally, species formation in another place – where a population on the periphery of the main population, but isolated from it, could easily and rapidly be swamped by a radical yet adaptive modification to the genome. But it was not until later that they realised how truly radical their idea was. Starting in that hospital bed, it was Steve Gould who began to realise that they had changed the face of evolutionary palaeontology for ever.

And that is why the palaeontology laboratory at the University of Bristol acquired its special role in the history of palaeontology. In the mid-1970s, Bob Savage's new graduate student, Peter Williamson, decided that he had the perfect natural laboratory to test the idea. The laboratory was in an area that Bob knew well, in the Lake Turkana region of the African Rift Valley, an area of outstanding yet hostile beauty where Bob and his beloved wife, the noted vertebrate palaeontologist Shirley Coryndon, had already worked for many years in their quest to recover the bones of Miocene carnivores. The Turkana region was already a legendary hunting-ground for palaeontology for it was where Louis Leakey and his wife had already found hominid fossils.

Like many a graduate student before and since, Pete Williamson had correctly identified the burning issue of the day and he pursued it with all the vigour and ambition that a hungry young scientist could bring to bear. He went to Turkana to try to find proof of the punctuated equilibrium theory. He did not work on vertebrates – Bob's speciality – for, like the dinosaurs of Wyoming, their fossils were too scarce. Instead, he worked on fossil molluscs – gastropods – that were deposited in long and apparently unbroken sedimentary sequences at the margins of the lake itself. Williamson and Savage speculated that the Turkana mollusc populations would be an ideal testing-ground for the Eldredge and Gould hypothesis, because the population was small and perhaps susceptible to allopatry (that is, at certain times in the past they could have been cut off from the main body of the population and able to evolve away from it).

They were right.

After several field seasons and careful laboratory work counting specimens and mapping morphological change among the population members, Williamson and Savage concluded that the Turkana gastropods did indeed evolve in rapid spurts in peripheral populations. They traced the origin of these spurts to occasions when the lake had dried up, killing off the bulk of the population in the main body of the lake, and putting the few that remained under intense selective pressure to come up with adaptations that would allow them to survive. Williamson published his paper in the prestigious journal *Nature* in 1981 and concluded that there were indeed long periods of stasis and short bursts of rapid species production, but there was no gradualism. So 'punk eek' had its first big success in fossilised mud deposits on the edge of an ancient lake in East Africa. After his doctoral work, Williamson went to Harvard to work with Gould. I still remember seeing his office door just yards away from the guru's himself. Williamson had made it. Big time.

Cake eaters and hoarders

The theory of punctuated equilibrium comprised one of the two major controversies which gripped the profession in the 1970s. (The other was cladistics, of which more later.) In the aftermath of Eldredge and Gould's 1972 paper which introduced the concept of punctuated equilibrium, palaeontologists were waking up to the knowledge that since the time of Darwin himself they had successfully blinkered themselves to the true nature of the fossil record. What is more, and they knew it, they had manifestly failed to be properly scientific. By denying the adequacy of the fossil record, they had effectively brainwashed themselves into overlooking the very thing they should have been looking for: the pattern of evolution as documented in the fossil record. Once the message had sunk in, a paradigm shift in the true sense swept outwards from America's East Coast palaeontological intelligentsia. It found a sympathetic hearing in the great schools of palaeontology in Johns Hopkins and Chicago, and sweeping onwards found a niche at the Woods Hole Oceanographic Institution where a young Swede named Bjorn Malmgren was working.

Chapter 2 discussed the importance of the planktonic foraminifera in stratigraphy. Many of the same characteristics that make them so useful in slicing time into useable segments also make them of value in evolutionary studies. In short, planktonic forams evolve 'easily'. In theory there are some important additional considerations that on the face of it should make the planktonic foraminifera the premier fossil group for studying evolution. Their small size should mean that thousands can be collected from a single sediment sample, and such samples should be able to be collected very close together in a rock section, providing an ideal opportunity to see evolution happening in the fossil record. Finally, their relatively simple shape means that it is possible to describe them mathematically by a technique known as morphometry. In practice, however, it was only since the seventies that these considerations could be properly exploited.

Morphometry has been around for a good chunk of this century and has been applied to a variety of invertebrate fossils. One problem however has always been that the data collection was labour intensive, using a ruler and protractor in the case of the larger invertebrates or a graticule attached to a microscope lens in the case of microfossils. Additionally, before the Deep Sea Drilling Project and the availability of huge numbers of well-preserved samples from microfossil oozes the number of specimens available for this type of analysis was often, after all, inadequate for statistical treatment. But the combination of the Deep Sea Drilling Project (DSDP) and the advent of digital computers changed all that. Suddenly, there were millions of specimens to work with and fast computers to collect and process the data. One of the most fundamental advances was teaming up shape detection software with appropriate video hardware to mathematically describe complex shapes.

The fact that DSDP samples can be collected continuously from records that span millions of years in uninterrupted sequence, and that the planktonic foraminifera are vastly superior in numbers to any other plankton group (with the sole exception of nannofossils) meant that the planktonic foraminifera became *the* group of choice for morphometrists in the 1970s to study the rate of evolution.

Bjorn Malmgen was among the first to recognise this and after an initial study that was not computer-oriented teamed up with Bill Berggren – the guru then as now of planktonic foraminiferal micropalaeontology and the brilliant Pat Lohmann. (The Lohmanns are legendary within the paleontological world. Pat and his brother K.C. are both fearsomely intelligent numerical and geochemical palaeontologists.) Pat was the man who invented 'eigenshape' analysis, the technique that would be most widely applied to the evolution of the planktonic foraminifera.

The statistics that are most commonly used in biology and the biological sciences seek to determine the degree of relationship between two features of an organism. Let us take one of the simplest possible examples, the relationship between body length and weight in, say, a group of fish. As size increases so too will body weight. The tightness of this relationship (i.e. how closely the data cluster around a line drawn through the cloud of data points) is called the degree of correlation.

The higher the degree of correlation – the maximum is 1 – the better the relationship. When the two variables both increase together (as is the case between two obviously age-related parameters such as length and weight) the correlation is a positive correlation. On other occasions, one variable can increase while another highly correlated variable decreases. This is called a negative correlation and the maximum negative correlation is –1. The slope of the line is also important for it denotes how rapidly one variable increases (or decreases in the case of a negative correlation) with another.

There is no *a priori* reason why the investigation of relationships between variables should be limited to two axes, however. In fact, in complex systems (such as living things) there may be no obvious relationship between the variables that are easy to measure. Groups of variables may work together in subtle concert and not therefore be visible to these ordinary 'two-dimensional' statistics. Multivariate statistics are a group of statistical techniques that consider multiple variables simultaneously and seek to identify new variables that we may think of as 'composite variables'. In other words, multivariate statistics create new summary variables out of the original group. Two of the most common multivariate techniques are factor analysis and principal components analysis. The difference between the two is arcane and need not concern us here except to note that factor analysis makes some assumptions about the degree of correlation between the starting variables and principal components analysis does not.

Eigenshape analysis is a modification of principal components analysis (PCA). Like factor analysis and principal components analysis it seeks to reduce the complexity of a data set with several axes. Here, though, the original data matrix is simply the angles between straight-line segments that are fitted around the edge of a shape, in this case a planktonic foram shell. To produce an average foram shape several (or usually many) specimens are measured from each successive sediment horizon and the resulting data matrix is reduced using PCA. This produces nothing less than an average shape that describes all the animal's shape variability from a particular time horizon.

The image analysis software can then reconstruct this average species from this particular time horizon graphically, in effect creating a 'virtual foram'. The next step is to pile several of these virtual animals on top of each other to make a lineage that shows their changing shape through time. And quite suddenly we find that we have a statistical time machine for the investigation of evolution in the fossil record.

Eigenshape analysis was a breakthrough. It was nothing less than a device that allowed us to see evolution occurring in the fossil record. It was a technological miracle. The first lineage of planktonic foraminifera to undergo this mercilessly quantitative treatment was the *Globorotalia tumida* lineage of the Neogene.

The results were startling. The data clearly showed a trend of shape evolution which at first glance looked a lot like the phyletic mode that Simpson and the others had promulgated. But on closer examination the *tumida* study highlighted a problem. In the heyday of the synthetic theory of evolution there is no question that this trend would have been hailed as a perfect example of phyletic gradualism; it looked so perfectly similar to the *pattern* envisaged by the Trinity, yet the evolutionary transition to *tumida* from its ancestor occurred quickly, within about two million years, a speed that was on punctuated equilibrium timescales. The *tumida* study therefore combined features of both phyletic gradualism and punctuated equilibrium and the cat was among the pigeons. When Malmgren and his co-workers, Bill Berggren and Pat Lohmann published their paper they named their new phenomenon 'punctuated gradualism' to show that evolution is vastly more complicated than any one model of it can ever portray. We now suspect that evolution in different groups of organisms occurs through different processes. Microfossils from the deep ocean are all single-celled and reproduce by cloning or by a mixture of cloning and sexual reproduction (where sex cells combine genetic information from parents). This means that the mechanisms of gene exchange envisaged by Eldredge and Gould may not apply. Also oceanic microfossils exist in vast populations (with trillions of individuals) which are spread across oceans if not hemispheres. Under these circumstances the formation of isolated populations (allopatric speciation) is not likely, and the opposite mode of evolution based on sympatry (literally, evolution in the same place) is likely to predominate. And microfossils routinely show a lot of shape and size variation that are simply related to environmental change and may not reflect genetic or evolutionary change at all. Microfossils would seem not to be the universal testing ground for evolution that they were originally hoped to be.

Among more complex organisms – macroinvertebrates and vertebrates – punctuated equilibrium does seem to dominate. Species do seem to persist unaltered for long periods (as Peter Williamson and Bob Savage showed with the Lake Turkana molluscs) and then undergo rapid bursts of intense evolution where new species are produced and the old lineages become extinct. Change under this regime occurs so quickly that it is rarely observable in the fossil record, thereby explaining the old paradox of why evolution was so rarely seen by the old science of fossils.

By giving us the appropriate model with which to examine evolution in some groups, and stimulating us to acknowledge that there must be a plurality of evolutionary mechanisms, Steve Gould and Niles Eldredge both qualify as architects of eternity.

The don and the death of neo-Darwinism

The issue of punctuated equilibrium and the environmental control of the shape of microfossil clones brings us to a different but related subject: one that goes by the rather terrifying name of 'functional morphology'. In the mid-1960s one idea in palaeontology exercised the mind of a Cambridge don – Martin Rudwick – above all others. Why do animals and plants *look* the way they do? Regardless of which model of evolution you adhered to – and, when Rudwick was at Cambridge, the synthetic theory was at the peak of its dominance – it seemed to him that there should be some formal connection between structure and function in both organisms and in their component organs. What reason was there for organisms to have the structures that they had? Why, for example, in Rudwick's own favourite group, the brachiopods, were the jaws of the shell commonly zigzag rather than simply straight? What difference did it make? Rudwick thought about this question and gradually developed a dictum that is now famous in the trade – that function governs structure in living and fossil organisms. Rudwick founded the paleontological subdiscipline of functional morphology.

Rudwick reasoned that for natural selection to work function *had* to govern structure in an organism's morphology. The most famous example of this is the eye. Light-sensitive organs are common in the animal world; think of the eyes of insects, spiders, advanced cephalopods (such as squid and octopus), as well as the design possessed by the vertebrates. In order for each of these structures to perform its basic functions all of them need to conform to the laws of physics. Each has to focus light, convert the photons into an electrical signature and transmit this data to the brain. This requirement confers physical similarity in the sense that each has for example a lens. The details of the way that the data is collected and handled has only minor differences but in each case the parallelism of similar functional requirements has resulted in the evolution of analogous structures. Thus function governs structure and similar functional require-ments result in parallel evolution of similar structure even in different groups.

The wing is another good example. It is found in the insects, the reptiles (think of the pterodactyls) and the birds. In all three cases, analogous function has clearly resulted in analogous structure. But here we can go further and recognise a condition known as 'homology'. The wings of pterodactyls and birds are homologous because they are based on a similar basic arm design – a radius and an ulna, supporting a hand – inherited from a common ancestor. In fact, this basic design is found, variously modified, in all vertebrates. In whales and dolphins, the arms and hands are highly modified to form flippers; in apes, the arm and hand are elongated for the manipulation of objects. The arm and hand are homologous structures. The functional governance of structure results in parallel evolution in invertebrates as well. Rudwick realised that the zigzag gape of the brachiopod shell is there to increase the surface area of water passing over the animal's gills which filter out plankton for food. This zigzag arrangement is also found in molluscs for the same reason.

Rudwick's theory of functional morphology was eagerly seized upon by palaeontologists. Clearly there was order to the universe if the laws of physics governed the shape of animals and plants. Nowhere was the functional morphology idea more enthusiastically taken up than among the micropalaeontological community. The same reasons that made the planktonic forams attractive to those interested in shape and species evolution made them the ultimate vehicle to test Rudwick's theories. In 1969, a micropalaeontologist named Richard Cifelli pointed out the similarities in the types of morphology in the four radiations of the group, the early Cretaceous, the late Cretaceous, the Palaeogene and the Neogene.

There are effectively two types of planktonic foram – the globo-rotaliid and the globigerine; ones that look broadly like saucers and ones that look broadly like tennis balls. Some years later, in the early 1980s, a British micropalaeontologist began to question what the adaptive advantage of this duality was. Malcolm Hart of Plymouth Polytechnic wondered if there were some adaptive advantage to the saucers and the tennis balls which was responsible for keeping the range of possible morphologies so narrow. At that time it was known from early plankton tow studies (particularly those conducted by the Californian biologist Allan Bé) that the saucers tended to live deeper in the water column than the tennis balls.

Hart, using classic uniformitarian reasoning, extended this and suggested that the saucers had (for some unknown adaptive reason) always lived deeper in the water column than the tennis balls. He suggested further that each of the four radiations of the planktonic foraminifera started in surface waters with the 'tennis ball' morphology dominating and that these gradually evolved into deeper water, acquiring along the way the 'saucer' morphology.

This seemed a reasonable idea – although nobody really had a clue as to what the adaptive advantages of the planktonic foram's shell morphology actually was, and the detailed timing of the acquisition of these morphologies in each of the planktonic foram's four radiations had not been studied sufficiently to be sure that things happened in the sequence that Hart suggested. There the matter rested until the late 1970s and early 1980s when the increased availability of well-preserved material from the Deep Sea Drilling Project and the rapidly growing number of isotope labs around the world meant that a sizeable data-base of the isotopic characteristics of many species of planktonic foraminifera through the entire Cenozoic and late Cretaceous had begun to accumulate.

It was the existence of the vertical gradients of carbon and oxygen isotopes in the ocean that provided the means to test Hart's hypothesis. The oxygen isotope gradient (temperature decreases with increasing depth in the ocean) had been known about since the early years of stable isotopic research in the fifties. Emiliani himself had even tried to establish where the planktonic forams that he proposed to trace the growth and decay of ice sheets had lived in the water column. By the early eighties the vertical gradient of decreasing carbon-12 with depth in the ocean was also understood.

The idea of using these gradients to rank different planktonic foram species in the water column was resurrected by Bob Douglas and Sam Savin, the first of many successful micropalaeontologist/isotope geochemist collaborations. They analysed several different species of planktonic foraminifera in the same sample from several different time intervals between the Cretaceous and the present day. Their paper, with its approach of ranking different species at different geological horizons along

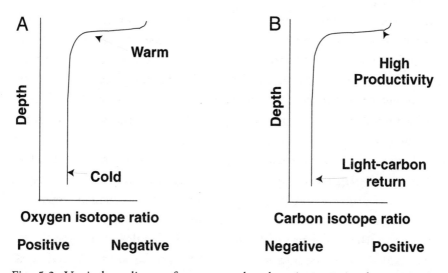

Fig. 5.2. Vertical gradients of oxygen and carbon isotopes in the ocean. A. Temperature decreases with depth. The shells of deeper-living planktonic foraminifera therefore have a more positive oxygen isotope ratio than those growing near the surface. B. Productivity decreases with depth so more of the light isotope of carbon is available for deep-living planktonic foraminifera to use in building their shells. This effect is compounded by the *addition* of light carbon at depth as dead animal and plant remains falling from the surface decay.

the vertical gradients of oxygen and carbon isotopes was both groundbreaking and in one vital respect fundamentally flawed. For Douglas and Savin's data apparently showed that the saucers lived – and had *always* lived – deeper in the water column than the tennis balls. But their analysis of Palaeocene material was based on only a handful of measurements. These data apparently showed that the tennis balls lived shallower than the saucers – as Hart had predicted – and in conformity with Rudwick's paradigm that equated similar structure with similar function.

Luckily, new data were to be forthcoming in the next few years. They were generated initially by Anne Boersma, a student of Bill Berggren's. Anne came to the Godwin lab – under Shackleton's reign by then one of the premier isotope labs in the world – in the late seventies to work specifically on the palaeobiology of the planktonics as it could be addressed using stable isotopes. The core material that Boersma studied in her sabbatical year came from one of the DSDP holes drilled in the North Pacific; DSDP 465. The north Pacific generally is an inauspicious place to search for well-preserved carbonate material. The waters here are old and therefore corrosively hostile to carbonate microfossils. However some of the sea-mounts of the north Pacific are relatively shallow and they poke through the calcite compensation depth like inverted dip-sticks. The sediments that sit on these are insulated – sometimes by only the narrowest of margins – from corrosion. So although it is rare to find Palaeogene sediments in the Pacific Ocean, when one does the preservation tends to be good. Additionally, the Pacific has been populated since the middle of the Cretaceous by a wide range of planktonic species. Boersma exploited the good core recovery and high species diversity together with the advances in mass spec technology that had been made in Shackleton's lab by Mike Hall to readdress the problem of the depth habitats of fossil planktonic foraminifera. Hall's preparation system was so good, in fact, that as few as five planktonic specimens could be reliably measured. Very quickly Anne realised the Palaeocene data that Douglas and Savin had published was unrepresentative. No matter how many times she and Mike Hall made the measurements on different samples and specimens the globigerinid (tennis-ball form) ranked consistently deeper than the globorotaliid (saucer-shaped) forms. They published their findings in the 'blue book' – the informal name for the monumental tomes that comprise the Initial Reports of the Deep Sea Drilling Project. The name 'Initial Reports' is misleading. It suggests that the work within is somehow quick and ephemeral and yet this is not so. For many years the 'blue book' was the receptacle for some of the most astounding work in the new science of fossils. Many scientists have devoted some of the best and most productive years of their life to writing the papers that went into the 'blue book' and in so doing helped develop the new science of fossils with its strands from palaeoclimatology, palaeoceanography, geochemistry and palaeomagnetism. Anne Boersma was one such. Eventually, though, Anne moved out of the subject and went to set up her own oil-company consultancy. But the enigma of the overturned planktonics remained.

DSDP hole 577 was drilled on the Shatsky Rise in the north-west Pacific in the early 1980s. From the perspective of the functional morphology of the planktonics it was an important site not only because it was near the Hess Rise Site (DSDP 465) but also because it had been drilled using new technology, the hydraulic piston corer, which pushes a smooth metal barrel into uncompacted sediment ahead of the rotating drill bit. Previously, the existing rotary core techniques had the effect of distorting the cores as they were drilled so that by the time they came up they'd been converted into something that looked a lot like that sad worm of toothpaste you find on the edge of the sink the morning after a heavy night when you forgot to put the top back on the tube. When the Palaeocene of 577 was retrieved, it had an almost perfectly preserved fauna of planktonics and benthics in the correct stratigraphic order. DSDP 577 has since become one of the most intensively studied DSDP holes in the world – not least because it contains an excellent Cretaceous–Tertiary boundary section. (In the mid-eighties such was the demand for K–T boundary material that the Deep Sea Drilling Project slapped a moratorium on those portions of cores that had the K–T boundary preserved in them in an attempt to protect some material. Only the luckiest, or best connected, investigators were able to get hold of samples from this crucial era boundary.) But at Shackleton's lab in the middle 1980s it was the upper portions of the Palaeocene that were being studied by, well, yours truly. This time it was not only different species that were measured. I had the idea of analysing different *growth stages* of these species too. Like ammonites, planktonic forams grow by accretion; new chambers are added over old chambers, but because of their small size it is not possible to successfully break off the chambers to analyse in a mass spectrometer. There is an alternative approach which is to sieve the dried sediment through a stack of soil sieves. Different sizes of foram are therefore sorted on the sieve meshes and can be picked out for subsequent isotopic analysis. This was a labour-intensive business; to measure the smallest of the growth stages of the forams required geometrically larger numbers with every one-step decrease in sieve size so that by the time I got down to what we called the 'j' fraction – a mesh size of 106–125 microns – I was having to pick about 300 specimens for analysis, an occupation that would easily consume a morning's work and leave me with white spots jumping in front of my eyes. Even then the whole procedure was a gamble. Occasionally, the machine would decide that there was not, after all, enough gas to measure and throw the sample away, an event heartbreakingly akin to flushing a Grand Cru Moët down the sink.

However, data was eventually forthcoming and it became clear that the smallest forms had very much less of the heavy isotope of carbon (carbon-13) than the bigger forms. This difference was very strongly size-dependent, with relatively more carbon-13 being present in each successively larger size fraction.

One explanation was that these forams were floating up in the water column as they approached maturity thereby capturing more of the heavy isotope of carbon (carbon-13) as they grew. The problem was that the oxygen isotopes did not show a comparably clear pattern of systematic *decrease* which would indicate the expected rise in temperature as the foram moved closer to surface waters. We were at a loss to explain the anomalous carbon isotope measurements, but one thing was very clear: no matter how many different species we measured throughout the Palaeocene and early Eocene, the globigerinid form – the tennis balls – always ranked deeper in the water column than the globorotaliid form – the saucers. The Hart hypothesis had collapsed.

There were some interesting rumbles in the jungle in the immediate aftermath of our publication of these data. It turned out that Fred Banner, emeritus professor at University College London was a closet isotope sceptic. Banner had worked closely with Walter Blow, the doyen of classical planktonic foraminiferal micropalaeontology in the fifties and sixties at British Petroleum where they were using planktonic forams to date oil-bearing strata. Water Blow was a real character. He had a wooden leg, a temperament unafraid of conflict and a passion for the planktonics that bordered on the obsessive. His monumental work *The Cainozoic Globigerinina*, completed and published posthumously, is a three-volume epic that summarised the planktonic foraminiferal discoveries of twenty years of BP's worldwide oil hunting. To call it densely written is to call *Ben Hur* simply 'a movie': like *Ben Hur* it is an epic. But an idiosyncratic epic. Blow's ideas on classification were outré to say the least and reflected his remorselessly pragmatic approach to biostratigraphy. On one notable occasion at which I was present as an awed graduate student, a world-famous micropalaeontologist of the succeeding generation (who shall remain nameless) was casting about for something with which to prop my microscope at a more propitious angle prior to advising me on some knotty problem of fossil identification. His gaze fell on Blow's book weighing down the end of my desk. Within minutes he had constructed a stand from the fabulously expensive and lovingly tooled tomes. 'Ah, a use for Walter at last . . .' he sighed.

When Blow died, Banner became BP's foram guru and eventually was appointed to the University College micropalaeontology chair while continuing to consult for BP. In the heady Thatcherite days of the mid-eighties very large oil companies like BP had spectacularly equipped research labs which included state-of-the-art mass spectrometers. The prospect of spiking Cambridge's guns was apparently too tempting to resist and Banner directed BP's isotope lab to find some Palaeocene planktonics and put these new ideas about the depth stratification of the planktonics – as promulgated by young Corfield – to the test. The project foundered when BP's own secret in-house measurements confirmed those of the Cambridge group. The old order had changed for ever – the functional morphology of even an apparently simple group like the planktonic foraminifera could not be taken for granted.

The Cladists

At about the same time as the revolution in evolutionary thought fuelled by Eldredge and Gould there was another revolution in evolutionary biology, mostly orchestrated by members of the Natural History Museum in London, and it was to do with how organisms were classified. In a way this too stemmed from dissatisfaction with the synthetic theory. For years animals and plants had been classified in a way that tried to reflect their evolutionary history but these theories were notoriously liable to change as the supposed evolutionary relationships between animals changed. Then there was the growing awareness that the fossil record would never supply all the missing links necessary to trace ancestry through uncountable generations. An evolutionary taxonomy was a chimera which could never be realised. Cladistics is an attempt to eschew reliance on an unknowable total evolutionary history. It makes no assumptions about evolutionary relationships but rather attempts to classify and show the relationships between organisms on the basis of objective criteria. Cladistics is an attempt to put taxonomy on a testable footing and therefore formalise it as a proper science. The basic idea of cladistics was put forward by a German entomologist, Willi Hennig, in his book *Phylogenetic Systematics* published in 1950 but it was not until the 1970s that Colin Patterson, Dick Jefferies and Pete Forey from London's Natural History Museum (known in those days as the BMNH) together with Brian Gardiner from the University of London decided to adopt the approach for the classification of fossils. They were stimulated by their desire to find the most likely candidate for the ancestor of the tetrapods (land-dwelling vertebrates). Since its earliest days, cladistics has been fenced about with some fearsome terminology that did nothing to promote its use. But we can grasp cladistics without recourse to such jaw-breakers by understanding that the guiding principle of cladistics is merely the search for shared, derived features. If we accept the idea that the outcome of evolution is the generation of new features – say, warm-bloodedness – then all of those animals that share warm-bloodedness belong in one natural grouping. Similarly, if we accept the idea that life on land has required the modification of lobe-fish fins into structures that we call arms and legs then all of the animals that share these features are a natural grouping. These 'natural groupings' are represented by nodes on a strictly regimented branching tree which is called a cladogram, and all the branches above each node represent a natural group, a clade.

Of course, the matter is more complicated than that and relies on the recognition of homologies (structures that share a common ancestry). Another complication is the fact that the greater the number of homologies that are incorporated into the database used to generate the cladogram the

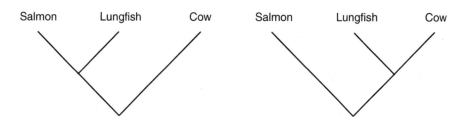

Fig. 5.3. Two cladograms illustrating the relationships between three animals: the salmon, the lungfish and the cow.

greater (often geometrically greater) is the number of possible solutions. Which one of these is correct? To decide between several competing cladograms – in other words, to find the right answer – the cladist uses the principle of parsimony; he chooses the simplest explanation. Of course, the simplest explanation is not always the right one, it is merely the most defensible.

Figure 5.3 shows two cladograms illustrating the relationships between three animals: the salmon, the lungfish and the cow. The diagram on the left shows the relationship between these animals as it was perceived in the classical tradition, stemming from the synthetic theory of evolution, namely that since the salmon and lungfish are both 'fish' they are more closely related than either is to the cow. The diagram on the right, however, illustrates the relationship as a cladist would see it, namely that less inference about possible (and probably unknowable) evolutionary histories is involved if the lungfish and the cow are grouped together since both have more shared, derived features in common than either does with the salmon. The fact that both systems can be portrayed by a cladogram testifies to the power of the technique. Only the latter is a true cladogram, though, for only it incorporates the cladistic principle of parsimony. The cladists preference for grouping a lungfish and a cow more closely together than a lungfish and a salmon incensed the traditional palaeontological establishment of the time. The cladists themselves delighted in the fact that, as they believed, only they had a truly objective scheme of classification. The dichotomy between the two competing salmon, lungfish and cow schemes came to be a metaphor for the split between the cladists and the rest of the palaeontological establishment.

In fact, the use of cladistics sparked a storm of controversy among the palaeontological old guard. Part of the problem was the terminology and the strange new relationships that cladistics could generate but a lot of the problem was the cladists themselves. They tended to form a rather exclusive clique that met in a pub near the Old Brompton Road (see Henry Gee's fine book *Deep Time* for more detail on cladistics and the cladists). From the perspective of the palaeontological community, the science of cladistics was fuelled by the bulk intake of lunchtime alcohol and its practitioners were much given to bellicose mutterings about the death of evolutionary theory. Colin Patterson referred to evolution on more than one occasion as 'a metaphysical research programme' and appeared on TV saying that he no longer found it necessary to believe in evolution at all! Of course this was perfectly true; cladistics makes no supposition about ancestors and descendants, it seeks merely to correlate related characters in different groups. But the way that it was said seemed calculated to send the message that the Simpsons, the Dobzhanskys, the Mayrs and all the other architects of the synthesis had made a mess of the whole thing and that it was only thanks to the refining intellects of the British Museum's staff scientists that taxonomy could now be called a science at all. The person who most actively took up the cudgels on behalf of the establishment was Bev Halstead, who wrote vociferous articles in the newspapers about the new display at the BMNH that was based on cladistics. He described the cladists as proponents of pseudo-Marxist ideology and did all but warn you not to let your children near them. This was good fun for the press of course, but I well remember the thirty-first symposium on Vertebrate Palaeontology and Comparative Anatomy in Bristol soon after I graduated. Bev and Colin were both there and out of view of the journalists they could not have been better friends. Wearying of the company of their fellow vertebrate palaeontologists they took the callow youth Corfield out to the most expensive curry house in Bristol where they both proceeded to indoctrinate him in the mysteries of industrial-strength inebriation. I was left with the indelible impression that the battle of cladistics was a put-up job for the TV crews.

Since then (both Bev and Colin have sadly died), the uproar has quietened down and cladistics is the most widely used classification technique in palaeontology. Of course, palaeontologists no longer deny the truth of evolution, but acknowledge cladistics for what it is, an objective tool that, like stable isotopes, has its role to play in the new science of fossils.

Heavy Metal and the Italian Rock Band

Boundary freaks

Science is as susceptible to changes in fashion as a high-street shopping junkie. Trends and fads come and go in palaeontology and the discipline's practitioners are as unable to resist a sexy new topic as any other scientist who wants to hit the big time. Nowhere is this more prevalent than in the way that palaeontologists relate to theories of the end, or as we call them in the trade, extinction models. And there is one particular type of palaeontologist who thrives on these extinction models – they are their *raison d'être* – and they are the boundary freaks.

Boundary freaks are a relatively new type of geologist who came into being really as a result of events set in motion in the late 1970s and early 1980s. But for palaeontologists generally, interest in extinctions is really as old as the science of fossils. It's hard to imagine it being otherwise. For extinctions – as we now know them – are what enabled the science's nineteenth-century practitioners to divide time up in the first place. For it was by recognising extinctions that palaeontologists arrived at what still are totally arbitrary divisions in the sedimentary record.

Why is this? It's because the rationale for palaeontology in the eighteenth and nineteenth century was the division of time (see Chapter 2). And the easiest way to divide time was on the basis of the most obvious and widespread markers in the fossil record. And these markers were also recognised in the most easily accessible places. In the nineteenth century, this pretty much meant Europe. And so, for example, the disjunction between the characteristic fossils of the Cretaceous and the Tertiary eras of Earth history were initially described here. For many years there was conflict about the placement of the boundary, and fossils thought to be Cretaceous were regularly placed in the Tertiary and vice versa. Eventually, the position, as defined on the basis of the invertebrate macrofossils, stabilised. But this was only a temporary fall-back position, for the boundaries needed to be identified in other ways if the original fossils that had been used to identify the boundary in the type area were lacking.

The concept of 'types' (as we saw in Chapter 2) is enormously important in palaeontology. But types are not restricted to the simple sanctification of a particular fossil specimen as being typical of a species. Palaeontologists also have type sections – rock sequences where the strata are undisturbed and therefore thought to represent an uninterrupted span of geological time. These type sections are then used as the standard against which all other rock sections thought to be of the same or similar age are compared. Then there are type areas where the rocks are designated as typifying those of a larger division of geological time. Type areas normally take their name from some local town or feature. For example, the type area of the Wenlock series which Elles and Wood studied is to be found outcropping near the tiny Shropshire town of Much Wenlock and the best exposures of Wenlock rock occur along a ten-mile-long ridge that runs south-west of the town, known as Wenlock Edge.

And so much of twentieth-century palaeontology has been spent trying to extend the boundaries to environments which the original scientists could not imagine, the deep-sea, perhaps. But can this useful but really rather dull enterprise of identifying sequences of sediments where fossils change sufficiently rapidly and noticeably to be used as signposts in the geological record explain the rise of boundary freaks in the 1980s? No, it cannot. The rise of the boundary freak was due to the sudden – some might even say explosive – surge of interest in the causes of faunal and floral change in the fossil record. And that particular rollercoaster can be traced to just two men and a rock sequence in central Italy.

The father, the son and the sleeping hills of Umbria, 43.25N, 12.32W. Sometime in the early 1980s

The hills above Gubbio slept as they always have, the occasional chirp of a cricket the only noise to invade the silence of the long road that winds between steepening cliffs of limestone. The Bottaccione Gorge is a very different place though to the Contessa Highway a mile to the east over the mountain. There the roads vibrate to the rattle of huge lorries transporting limestone from the quarries on the west side of the valley. And the Bottaccione does not have anything like the snake pit that the Contessa has. A deep, uninviting little gully that smells of cat pee and digs deep into the succession of strata just to the north of the K–T boundary. And so Walter Alvarez and his new team from the Renaissance Geology Group at Berkeley are grinding their way up the hill out of Gubbio along the Bottaccione Gorge to where the bar is. It's closed. No surprises there: the bar at the bend in the Bottaccione always seems to be closed these days. Perhaps the owner has retired on the profits of geotourism? A beer would have been nice, if only to break the journey before they introduce the old man to the clay layer. His father is in the passenger seat but he seems cool: Luis Alvarez is winding down, his volatility cooling with the passage of the years, like an old volcano. As a young man he was incandescence personified among the firebrand glitterati, nothing less than a physicist on the Manhattan Project. And then he went on to the work for which he won his Nobel prize in 1968, the discovery of several subatomic particles. But that's all behind him now. He's here to celebrate his swan-song, his final if perhaps not his greatest achievement. A piece of geological research that he would never have got into if his son had not decided all those years ago to become a geologist.

The jeep grinds further up the road, the cliffs of pink and white limestone slashed at regular intervals of a metre or less by narrow fibres of brown clay, technically known as marl. Eventually the road curves gently into a bend and there Walter pulls off, climbs out, stretches and looks around. He spreads out the map on the bonnet of the car. The silence is absolute, deep and eerie, punctuated only by the click of cooling metal. They aren't the first people to have been here. The gorge has been well known since the Italian geological survey mapped it in the aftermath of the Second World War. Then in the sixties a young micropalaeontologist from Milan, Isabella Premoli-Silva, logged the succession of microfossils along it and discovered an enigmatic clay band about half-way along the length of the gorge.

In the 1970s the gorge was host to the attentions of two of the brightest geologists of their generation – Pete Scholle and Mike Arthur – who showed that the particular era boundary represented by the clay layer is associated with a major upset in the carbon cycling of the oceans. But it is since 1980 that this place has become the most famous shrine to geological science in the world.

Walter remembers that for him this all started with something far more prosaic, even if it was a 'solid' piece of science. He had originally been interested in the palaeo-magnetism of the area but then became interested in measuring sedimentation rates – the speed with which particles of sediment float downwards from the surface of the ocean especially in the interval represented by the clay layer. This was the K–T boundary, the division between the Mesozoic and the Cenozoic. Sedimentation rates have historically always been a difficult thing to measure. Fundamentally it boils down to only two things – knowing where undisputed time markers are in the sequence of (often otherwise undistinctive) strata, and then knowing the thickness of the sediment between these two time lines. By calculating the difference you get the sedimentation rate. As Walter leans back against the jeep and stares down the hill towards the bend it occurs to him again that this is nothing less than the history of stratigraphy. To know time and thickness of rock is to know the nature of time itself – the fundamental essence of palaeontology.

But the way that they tackled the problem of trying to measure the sedimentation rate in a clay layer only a few centimetres thick – the one place in the Bottaccione Gorge where there were no forams to date the sequence – now that stemmed from the fertile, restless brain of his father. Luis' hypothesis had been: what if there were an independent measure of the rate at which sediments accumulated? Suppose, for example, it was not necessary to know where the time horizons in a rock sequence were. Suppose instead that your measure of time came from somewhere else entirely.

It was a physicist's solution. Unencumbered by a formal geological education Luis had a fundamentally different approach to the problem – a different 'take' if you will – and consequently saw things from a more flexible perspective.

Walter realises that now, realises that by inviting his dad to get involved with his own work, he has been responsible for a seismic shift in the history of palaeontology. Probably the biggest since Harold Urey unlocked the key to fossil climate research three decades before. Palaeontology has come out of the closet. Palaeontology now talks to other disciplines on an equal footing.

Luis is out of the jeep now, walking towards him. Walter leads him down the hill. The limestone ribs are interrupted here by a great trench dug in the cliff. This part of the succession used to be level with the rest. The trench is the result of the legion of geologists who have descended on this place since the mid-1970s. They excavated it so deep that you now have to lean far into the rock to find what you came for.

The rock around this area too looks as though it's been in a war. Everywhere it is pocked by perfectly circular holes an inch across, legacy of the palaeomagnetists and the pestilential rock-drills they use to collect their samples. So many palaeomag samples have been taken that in only five years this once pristine outcrop has been converted into something that looks disconcertingly like either a gruyere cheese, or maybe that car that Bonnie and Clyde met their maker in at the end of Peckinpah's film.

And then Luis leans in close and peers into the bottom of the trench. It's hard to locate in the high sunlight of the middle of the day but Walter points it out and eventually he sees it. There at the very bottom is a narrow band – only a couple of centimetres wide – of reddish brown clay. It comes up out of the ground at his feet, extends through the trench and continues on up the cliff face out of reach. The cause of all the excitement: the Cretaceous–Tertiary boundary clay. The famous outcrop of the K–T boundary in the Bottaccione Gorge where Luis and Walter Alvarez discovered the iridium anomaly which led to the asteroid impact theory of the extinction that killed the dinosaurs.

Dying for a living, boy?

The Cretaceous–Tertiary boundary is commonly abbreviated to 'K–T boundary'. The 'K' comes from the German 'Kreide' meaning chalk, for chalk, as we have seen, is a common sediment of the Cretaceous. The 'T' stands for Tertiary – the third era of life's history – although the term has largely been superseded by the more technically correct Cenozoic which encompasses both the Tertiary and the Quaternary eras. However the euphony of the term 'K–T boundary' resists all attempts at modification. It is now an inextricable part of the English lexicon and unlikely to change.

And yet it was not as though Walter had not already been working in the Bottaccione area for years. He and fellow post-doc Bill Lowrie, both of them from the Lamont–Docherty Geological Observatory of Columbia University in Ithaca, New York had contributed to some seminal work in this area. Like several other geologists in the seventies, they had become interested in the remarkable limestones of the Bottaccione Gorge. The limestones are remarkable because they were deposited in the deep sea – they were not shallow-water limestones like those that outcrop on Wenlock Edge. These limestones are different, deposited in about the deepest water you can get and still find fundamentally undissolved.

They soon discovered that another group, led by the legendary Al Fischer of Princeton as well as Mike Arthur and Isabella Premoli-Silva, were working on the very same question and by combining forces they produced a seminal memoir on the geology of the gorge. Originally it had been the happy conflation of decent foram stratigraphy and palaeomagnetics that had led geologists to the Bottaccione Gorge. But it was clay horizon at the K–T boundary that made them stay.

Walter talked the K–T time-duration problem over with his dad and it wasn't long before the famous Nobel laureate had arrived at a solution. They would use Beryllium-10, an isotope that was formed in the upper atmosphere when incoming cosmic rays collided with oxygen and nitrogen atoms. The half-life (the time needed for half the original quantity to decay to another isotope) for Be-10 had been calculated as 2.5 million years. There would be enough Be-10 still left in the clay for them to measure what the original concentration must have been (assuming the rate of production in the upper atmosphere was constant). And then the whole concept came unglued – the half-life of Be-10 was recalculated and found to be only 1.5 million years – and that one-million-year difference was crucial. There would not now be enough Be-10 left in the clay for them to measure. The idea was abandoned.

But around about this time Walter was appointed to Berkeley where his colleague Rich Muller and his father were already on the faculty. The problem of the duration of the K–T boundary would simply not go away. And once again it was the father who had the idea, which was related to the Be-10 concept yet different. It was based on another uncommon element – iridium, a member of the platinum group of elements. Iridium is in short supply on the Earth's surface. It has sunk into the deeper layers of our planet, the mantle and the core. A heavy metal, indeed. But, outside the confines of our own planet, iridium is in much greater supply as it is one of those primordial elements that was flung far and wide after the birth of the universe. Could significant quantities be dropping in a steady rain on to the earth? If so, the amount of iridium in the clay layer was at best going to be in the parts per billion (an American billion) range. To measure this small a quantity, cutting-edge techniques would be required. But on the Berkeley campus there was one man who they knew could help: Frank Asaro.

Asaro listened cordially when they diffidently approached him and then dropped a bombshell. He and a colleague were already working on something very similar, although their emphasis was on measuring the duration of deposition of fossil soils. However, if his co-author agreed, Asaro would be willing to make the measurements for the Alvarez team. A deal was struck and Asaro agreed to help. The technique Asaro was using to count iridium atoms was neutron activation analysis – and its minimum analytical requirement was nothing less than a nuclear reactor.

Neutron activation analysis works by irradiating samples. Neutrons from the reactor collide with the atoms in the sample and stimulate them to emit gamma rays. The electromagnetic spectrum of the resulting ray is as characteristic as a fingerprint and allows the various elements within the sample to be measured at levels down to parts-per-billion (ppb). The technique takes time however if the element being searched for is rare. It is not uncommon for samples to be irradiated for several months.

In this case the wait was even longer than usual for the universal scientific bogeyman – mechanical failure – had stopped by Frank's lab and killed his machine. It was several months before it was back on line, and not until the early summer of 1978 when Walter got the summons from his dad, 'Frank's got the data!' They attended, full of expectation – and their hopes were dashed. Their calculation was that at most, if the time represented by the Gubbio clay across the K–T boundary was their worst-case scenario of a few thousand years, they would expect to find 0.01 ppb of iridium in the rocks. Instead they found 3ppb – three hundred times the amount that they had expected. And after Asaro had realised that a mistake had been made in the chemical preparation of the sample, they discovered that they actually had 9ppb in the sample – nine hundred times the amount that they were predicting.

At moments like these in the life of a scientist your career boils down to two alternatives: either you have long drawn-out hair-tearing sessions in your office until your partner threatens divorce or separation, or you have long drawn-out drinking sessions in a bar someplace which continue until your partner threatens divorce or separation. The Alvarez team had to explain this high concentration of iridium. Either the iridium really did come from meteoritic dust and some extraterrestrial explanation was required, or there was some wrinkle in the way that iridium was deposited in sea water that nobody had yet thought of and the whole deal was a busted flush. To test this it was essential to try again, with another sample. Walter searched the literature for another complete K–T boundary section. Denmark was the only other obvious candidate.

I've already mentioned the bar which nestles under the cliff halfway between Gubbio and the K–T outcrop in the Bottaccione Gorge. Yet this bar, even if you can find it when it's open, is not *the* bar. For the nascent boundary freak there is but one true bar, and that is the one true bar above the one true section, at Stevns Klint on the eastern coast of Jutland in the tiny hamlet of Hojerup. The bar above the cliff at Stevns Klint is a remarkable and supremely civilised place: a rambling, gabled building built in the Scandinavian summer-house style on stilts, whose wooden floor protrudes far out over the edge of the gently eroding chalk cliff that faces the strait between Jutland and Zeeland. The light Scandinavian beer here is as good as anything that you will find anywhere. The Carlsberg is a work of art. But there is also something eerie about the village of Hojerup. There's a stillness to the place as though it is somehow aware of the secret it guards. Go there in summer and this small Danish village slumbers in a limpid summer sunlight that you wouldn't ordinarily associate with the Baltic states. Go there in winter and the rain falls in whispering sheets out across the strait towards southern Sweden. It's so quiet that after a few moments you can hear your own heart beating – a few more minutes and you can hear somebody else's – but there's nobody there. No one except the solitary barman quietly polishing glasses in the otherwise deserted bar and the hum of the chiller cooling the Carlsberg.

The Bottaccione Gorge has a similar quality, a place strangely out of time, where the clay layer at the K–T boundary sleeps between its enfolding limestone ribs. For anyone with any knowledge of palaeontology and any form of empathy these are places of ultimate endings – and beginnings.

Walter must have felt the same thing when he made his first visit to Stevns Klint in 1978. The mission was critical – they knew that if they could not find the iridium anomaly in the Stevns Klint section then they could not prove that the iridium anomaly was at least regional (they were hoping of course that it was global). And in that case, even if they replicated the measurements from Gubbio and came up with the same answer then the likelihood was that the iridium anomaly at Gubbio would turn out to be no more than some strange artefact of sea-water chemistry that might be worth some short article in *Geochimica et Cosmochimica Acta*, but would not be the stuff that *Science* articles are made of.

The Stevns Klint material did not look at all like the Bottaccione material. This layer was thicker and blacker and the surrounding rock was friable chalk, not hard limestone alternating with marl bands. This was no surprise to the Alvarez team since any time-equivalent material from the geological column can be expected to have a different aspect according to the different places that it is found. The rock type depends to a large extent on the environment that it was originally laid down in – the facies or environment of deposition – which will vary according to whether the sediment was laid down in the deep ocean, a shallow sea or on land. Also, sediments are almost always more or less modified from their original state by the many different diagenetic processes that have occurred during their long passage across the millennia.

The samples were duly collected and taken back to Berkeley for analysis and the iridium anomaly was found once again – the Berkeley group had the proof they needed. In so far as they had good evidence that there was an increase in extraterrestrial material at the K–T boundary they were ready to publish, but they lacked a detailed mechanism, a detailed description of the killer. In 1971 two palaeontologists, Dale Russell and Wallace Tucker, had suggested that the extinction at the K–T boundary may have been caused by a supernova near the Earth. Until the Berkeley group's work, the idea had remained entirely hypothetical. But when the iridium anomaly was discovered and confirmed at Stevns Klint then independent evidence for an extraterrestrial cause was at last to hand. Luis suggested an independent test for the supernova hypothesis. When the collapsed star recoiled and threw iridium out across the universe it would also throw out plutonium which could be measured in the boundary clay in the form of the isotope Pu-244. Careful analysis revealed – disappointment. Despite some initially promising data there was not enough Pu-244 in the boundary clays from Bottaccione and Stevns Klint. The supernova hypothesis was dead.

The only other credible idea was that the killer was an asteroid which had hit Earth bringing with it sufficient iridium to create the anomaly. But the details of the killing *mechanism* would not stand up to scrutiny. How could a simple impact, no matter how large, kill off a large percentage of the life on Earth? There was no easy answer. Walter returned, dispirited, to his palaeomagnetic work in Italy. In California however, after a lifetime of pugnacious tenacity, Luis was not ready to give up. He remembered an encounter he had had with an obscure Royal Society publication on the eruption of Krakatoa and the dust pall that had surrounded the globe after the detonation of that corked volcano. And so the idea of the K–T winter was born. A dust cloud surrounded the world in the immediate aftermath of the K–T impact event and decreased light intensity to a level where photosynthesis on land and in the oceans was halted. Cut off the fuel and the engine will die – and photosynthesis is ultimately the fuel that powers the world.

The K–T boundary and the death of the dinosaurs became front-page news. The old adage of uniformitarianism was finally overturned by the resurgence of the old notion of catastrophes as major turning points in Earth's history. And this time the architects of eternity were an ambitious young geologist, a physicist and a couple of chemists.

This was a major turning point. The old order – with its unhealthy preoccupation with gradualistic notions – was finally defeated, going over the next several years not gently (in fact fighting tooth and nail), but going nonetheless, into this good night. The hard realities of an outside scientific world where analytical technology and numeracy reigned supreme had scaled yet another of classical palaeontology's last bastions.

The palaeontological establishment split rapidly into several camps: those who could not accept the idea on any grounds, particularly the older generation whose uniformitarian roots were too deep to shift; those who did not like the idea because they found the evidence unconvincing; and those who loved the idea because it was novel and had a dangerous frisson of cataclysm about it.

In the aftermath of the Alvarez paper, it became clear that they had succeeded in priority of publication only by the skin of their teeth. A

Dutchman, Jan Smit, was on their tail with his own story of noble metal enrichment in the K–T section at Caravaca in Spain and almost simultaneously other groups confirmed the iridium enrichments in Denmark and New Zealand.

There was in those days – and to an extent there still is, at least among certain die-hard sectors of the palaeontological community – a feeling, so deep that it is never spoken of, that palaeontology should only be practised by those who disdain the vulgar requirements of high technology. These, of course, are the practitioners of the old palaeontology. Luis Alvarez bounded into the fray with enthusiasm. He and Walter fought hard to gain acceptance of the theory among the reactionary old guard, yet really never made much progress, their success being with the younger members of the community. It was they who went to seek out other boundary sections in a variety of different depositional environments. Early notable finds by these new acolytes of apocalypse were Carl Orth and Chuck Pillmore who together found several sections in a non-marine depositional environment in New Mexico interpreted as a fossilised coal swamp.

One healthy development that arose very quickly after the new era dawned was the Snowbird Conferences. The first was in the early eighties in Snowbird, Utah (hence the name). The conferences were designed to be interdisciplinary and to allow astronomers, astrophysicists, physicists, chemists, biologists and palaeontologists to learn each other's language. During the eighties the K–T community – and by about 1985 it seemed that almost every university geology department in the Western world had at least one researcher working on the boundary, such was its importance – was focused on finding the impact structure itself, what Walter Alvarez has called the 'Crater of Doom'. (See Walter's book *T. Rex and the Crater of Doom* (Penguin, 1998) for the full details of the K–T story and the hunt for the impact structure.)

But the camp of what we may call the scientific unbelievers had found a suspect of their own which could also account for the iridium anomaly.

The best bang since the big one

Iridium may be a material of the solar system, but it is also a material of the inner earth – the mantle and the core. It was too heavy to hang around on Earth's crust as it cooled and stabilised and instead sank into the depths of our planet. Quite soon after the Alvarez paper had come out, a group of geologists in the States realised this and understood further that the iridium enrichment could therefore be interpreted quite differently. On the face of it, the Berkeley group's asteroid theory seemed plausible, but the iridium could also have risen to Earth's surface via volcanic vents. Dewey McLean at Virginia Tech was among the first to suggest that the iridium anomaly at the K–T boundary could have been caused by volcanic activity and then the idea was taken up and championed by Chuck Officer and Charles Drake of Dartmouth College, New Hampshire. These three proposed that the iridium had come from the centre of our own planet via volcanism – and lots of it.

The west of India is covered by a fossilised sea of basalt. This is an igneous rock, which is to say it is one of a group of rocks that are formed by material coming from the mantle and core. And this igneous rock is of a singular age. On the basis of the existing radiometric dates, the Deccan Traps were known to be approximately the same age as the K–T boundary. Officer and Drake reasoned that the volcanism that gave rise to this enormous flood of basalt (covering an area as large as France), would have been sufficient to bring enough excess iridium up from the Earth's core to account for the iridium anomaly at the K–T boundary in the Bottaccione Gorge (and at the other sections around the world where it was rapidly being found as the asteroid impact scenario gained momentum). As radiometric dating of the Deccan Traps continued, pursued most energetically by the French geophysicist Vincent Courtillot, it became clearer and clearer that this remnant of an extreme volcanic episode was indeed contemporaneous with the K–T event.

By the mid-eighties the development of a palaeomagnetic reversal stratigraphy (*see* Chapter 3) had extended all the way back into the Mesozoic and was well-integrated with absolute dates as well as global biostratigraphic datums based on planktonic forams and calcareous nannofossils. This integrated scheme is known formally as the 'Geo-magnetic Polarity Timescale' but is usually (and mercifully) shortened to GPTS.

Normal and reversed intervals are given numbers. The K–T boundary occurs within reversed polarity interval 29 – or Chron 29R. Courtillot's dating of the Deccan Traps (the word 'trap' by the way comes from the Scandinavian term *trappa*, which means stairs) showed that they spanned three polarity intervals – from the top of Chron 30N to the beginning of Chron 29N, encompassing Chron 29R and corresponding to an interval of somewhat less than a million years. This was strong evidence in favour of the volcanism hypothesis which was particularly strongly championed in France, in part perhaps, because of Courtillot's high profile among the French scientific-political establishment.

Where did this new support for a volcanic cause for the K–T extinctions leave the Alvarez scenario? The problem was that the Berkeley group were continuing to be embarrassed by one huge missing piece of evidence – the impact crater itself. There were some early red herrings – the Manson crater in Iowa was investigated but found to be the wrong age. Then the search turned to the ocean. Tiny spherules of a mineral called sanidine had been found in K–T boundary sections in Spain and Italy and a young Italian geochemist called Alessandro Montanari, who had gone to Berkeley to work with the Alvarezes for his PhD, deduced that this was the alteration product of the original minerals olivine, pyroxene and calcium-feldspar. This suite of minerals is characteristic of oceanic, not continental crust. The Alvarez group, on the basis of this geochemical evidence turned their attention to the ocean. They knew that there was a good chance (about 25 per cent) that this search would be fruitless because the sea floor is being continuously consumed at the continental margins in areas known as subduction zones. And so, very early on, the Alvarez team made an assumption that the crater had been subducted and stopped looking for it because they believed that it was no longer there. This in turn meant that for all of the eighties the question of volcanism versus impact went unanswered.

Dr Strangelove, I presume?

Despite the non-appearance of the crater by the mid-1980s, the hold that the Alvarez hypothesis exerted on the palaeontological community was colossal. All around the world research efforts were being rejigged so fast you could practically hear the brakes locking. In Zurich one senior scientist, Ken Hsu, who had a well-tuned instinct for the sensational in palaeontological science, went all out for the impact hypothesis. His restless mind took Luis' original idea and elaborated it. Hsu was one of the early pioneers of investigating the way that carbon isotope ratios changed at the K–T boundary. The early work of Scholle and Arthur in the seventies on the limestones of the Bottaccione Gorge had shown that the ratio of carbon isotopes (an indicator of oceanic productivity) shows a dramatic inflection to more negative values at the level of the K–T boundary.

Although the measurements in the Bottaccione Gorge had been performed on limestones, other measurements, particularly in Deep Sea Drilling Project and Ocean Drilling Program (ODP) cores, confirmed this carbon isotope 'excursion' all around the world. Measurements of foraminifera, not just undifferentiated oozes, showed the same pattern of a sharp decline in the ratio of carbon-13 to carbon-12 at the level of the K–T boundary. Detailed analyses showed that the magnitude of this blip in benthic forams was much less than in planktonic foraminifera and suggested that the decrease was differently recorded in surface and deeper waters. It was Ken Hsu's fertile imagination that supplied an explanation for this difference. Inspired by the image of a global winter caused by a dust cloud wrapping the planet, Hsu immediately saw that this would decrease the speed with which the light isotope of carbon was consumed by photosynthesic algae living in the surface waters of the ocean. An immediate consequence of this would be that isotope records generated from planktonic foraminifera would show a greater negative inflection than records from benthic foraminifera (which form their shells away from areas where sunlight drives photosynthesis and depletes the water in the light isotope, carbon-12). Hsu named this effect, with characteristic panache, the 'Strangelove Ocean'. The Strangelove Ocean was an ocean whose surface had died – no living thing remained in it to pump light carbon out of the surface waters and transport it deeper. And if the oceanic plankton were dead, then it followed that everything else that lived in the ocean and which ultimately depended on plankton for survival would soon die too – within a few months or a few years at the most. The concept of the Strangelove Ocean was very soon modified into an apocalyptic scenario for our own age as well, and taken up and championed by none other than Carl Sagan. For the Strangelove Ocean was a concept of its time, and that time was the era of Ronald Reagan and his Strategic Defense Initiative (SDI, or the 'Star Wars' programme).

Sagan took the idea of the Strangelove Ocean and ran with it. Big time. By the mid-eighties he was one of the most respected scientific commentators in the US and commanded attention across the whole communication spectrum from popular TV to the Senate. It was known from the Alvarezes back-of-the-envelope calculations that the size of the impact (*assuming that it had occurred*) would have exceeded the combined megatonnage of all the nuclear arsenals in the world and this immediately begged a question: what would happen if there were to be a nuclear war? Could it trigger another K–T type winter? Could the *entire planet* be imperilled by dust-induced cooling? Could a *local* nuclear exchange trigger off dust-induced cooling? In fact, just how much of a nuclear war would be needed to start the big chill? These and related ideas dropped like bombshells into the febrile atmosphere of Ronald Raygun's America. The idea took terrifying hold of the public imagination. Suddenly there was no such thing as a limited atomic war any more and a new term – 'nuclear winter' – entered the testosterone-charged lexicon of the nuclear vernacular. It was just the excuse that Ronald needed and the SDI bill was passed and immediately work started across the US on the various components that would be needed to expedite the President's vision: giant space-based mirrors, anti-ballistic missiles, anti-anti-ballistic missiles, X-ray, free electron and excimer lasers . . .

So just who was it who said that palaeontology is an obscure subdiscipline of biology that will never change the world?

The metronome of spoliation

As if the nuclear winter and the Strangelove Ocean were not enough, in the spring of 1984 the stakes were upped once again. By now every palaeontologist in the world knew that if they hit the right buttons they would be taken seriously at the very highest levels of government, and more importantly that the coffer-dams on the funding agencies would swing open and tidal waves of money would come roaring out. The Alvarezes had done no less than create palaeontology's very own Manhattan Project.

There were two men with just the ticket. David Raup and J. John 'Jack' Sepkoski were members of that rarest of breeds – fully numerate palaeontologists. Raup had been one of the authors of the most innovative and influential textbook on the subject (Raup and Stanley, *Principles of Palaeontology*, W. H. Freeman, 1971), a book that argued for the use of rigorous statistical techniques in the interpretation of palaeontological data and looked at the fossil record from the perspective of fossils as once-living organisms. And Jack Sepkoski was the man who had one of the most important palaeontological data sets in the world. Sepkoski had compiled data on the time ranges and diversity of fossil taxa down to the family level for the entire span of the Phanerozoic, the last 570 million years. Jack Sepkoski had a unique perspective on the history of life. Jack saw the ultimate big picture. In his charts and compilations and computerised summaries he, for the first time, saw creation spread out before him.

Jack Sepkoski was a tall man with a terrific walrus moustache and a long mane of hair often tied back into a pony-tail who looked precisely what he was – one of the young minds from Harvard's influential palaeontological school (intellectual home of none other than Stephen Jay Gould) who would set the world alight. Jack's character was electrifying, too – dominant, ebullient, overwhelming, hard-drinking, he was the Ernest Hemingway of palaeontology, as well as being one of the finest, kindest men in the field. Jack once said that as he sat and typed in his summerhouse in Rhode Island, he would customarily measure the worth of his papers in bottles of bourbon. A one-bottle paper was OK, say *Journal of Paleontology* material, a two-bottle paper might be *Proceedings of the National Academy*, a three-bottle paper was the ultimate though: *Science* or *Nature* stuff. No doubt it was an exaggeration but it was characteristic of the man – he was a breath of fresh air everywhere he went.

The techniques that Jack was applying to his data set were some fairly arcane statistics in order to see if there was any underlying pattern in the data. There was such a pattern, particularly in the sequence of mass extinctions since the end of the Permian (the geological period that ends the Palaeozoic era), and it seemed to point to an underlying regularity in the spacing of rates of extinction. Or, to put it another way, there were peaks in the background rate of extinction with a period of 26 million years.

Raup and Sepkoski realised immediately that regular spacing of

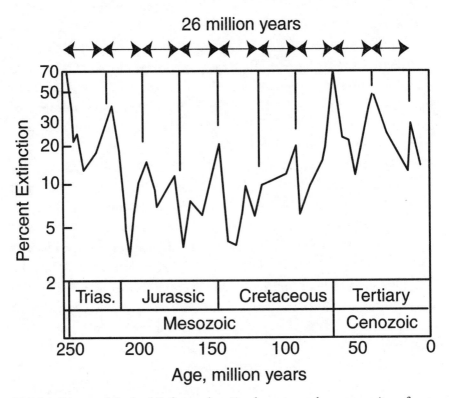

Fig. 6.1. Raup and Sepkoski's figure showing the apparently even spacing of mass extinctions, with a cyclicity of 26 million years, over the past 250 million years, since the major extinction at the Permian–Triassic boundary.

extinctions implied a common, cyclical cause and that this was most likely to be found in the extraterrestrial realm where the planets and stars dance to the ceaseless tune of orbital mechanics. They chose a couple of high-profile meetings to float their results at and made sure that they circulated pre-prints of their paper to every astronomer and astrophysicist that they could think of.

The resulting furore resembled one of those piranha feeding frenzies that occurs when the bad guy from SMERSH drops the trap door in the floor of the lift in the second reel of the Bond movie. Several groups of astronomers and physicists immediately took themselves off and sat down to work out an extraterrestrial killing mechanism. Two of them thought that the answer had to lie with Earth's vertical movement in the galactic plane. The period of this cycle is approximately 60 million years – that is the time taken for the sun to move north from the galactic plane, halt, then move southwards, through the debris, planets and stars that comprise the galactic disc to its southernmost extremity then return to the galactic disc. This was the essential attraction of the idea for the astronomers; it had to conform – at least within statistical error – with the estimate of extinction periodicity that Raup and Sepkoski had calculated. However there was still, as a suite of papers in the journal *Nature* amply demonstrated, plenty of room for imaginative thinking.

Richard Schwartz and Philip James of the University of Missouri relied on estimates made by the astronomers Oort and Bahcall that the half-period of the sun's oscillation through the galactic plane was between 34 and 31 million years. They decided that the errors in the Raup and Sepkoski estimates were such that this was still in good agreement with the 26-million-year period in the extinction record. Their preferred killing mechanism was an increase in the flux of X-rays and cosmic rays every 26–31 million years as the sun and solar system reached maximum distance away from the galactic plane. Since a prerequisite for any cyclical extinction mechanism was uniformity of killing agent, Schwartz and James, via some bare-faced, scientific special pleading, decided that the extinction at the K–T boundary – which had started the whole debate – was a one off, a special case. Sure, an asteroid hit the planet 65 million years ago, but this was the exception rather than the rule. Periodic extinctions were caused by radiation, increasing in intensity to lethal levels every 26 million years when Earth was far away from its galactic neighbours.

Rampino and Stothers in the same issue of *Nature* re-computed the Raup and Sepkoski figures and disputed their estimate of 26 million years. Their estimate was about 30 million years, which was close to the original figures of Oort and Bahcall. More importantly, and unlike Schwartz and James, Rampino and Stothers' explanation included the K–T boundary event. They suggested that the killing agent came into action as the solar system moved *through* the galactic disc itself (Schwartz and James' mechanism came into operation when the solar system was furthest away from the galactic disc). Their hypothesis was that as the sun moved through the galactic disc it gravitationally disturbed comets and asteroids orbiting far out in the cold reaches of the solar system impelling them to cascade inwards towards the system's centre and Earth. These impacts brought with them the extraterrestrial iridium that had provided the anomaly at the K–T boundary. Like all good scientific ideas, the Rampino and Stothers hypothesis made some definite testable predictions of its own, namely that every mass extinction in the fossil record should have an iridium anomaly associated with it.

Two other papers in that issue of *Nature* dealt with a different type of killing mechanism, but within that both were very similar indeed. The first, by physicists Daniel Whitmire and Albert Jackson, was encouragingly entitled 'Are periodic mass extinctions driven by a distant solar companion?' but was followed by a piece so impenetrably written as to seem almost incomprehensible to everyone but another astrophysicist. Their idea was that the sun had a distant solar companion, which had remained invisible possibly because it was either a black hole or a brown dwarf. The idea was that this unknown solar companion – perhaps another star – revisited the earth every 26 million years or so, dislodging a comet or two from the Oort Cloud which hit Earth and caused mass extinctions. In the same issue Davis, Hutt and Muller published a similar hypothesis entitled 'Extinction of species by periodic comet showers'.

The *Nature* editor of the time, John Maddox, devoted an entire page at the front of the issue to a pointed complaint about the timing of these papers. He and the journal clearly suspected some sort of stitch-up but that didn't stop them publishing the most important paper of that quintet. Muller and Alvarez's paper in that seminal issue of *Nature* provided the invaluable corroborative evidence. Muller and Alvarez had compiled the ages of the craters from the past 250 million years of Earth's history and found a periodicity of about 28 million years in the data – close enough to be statistically congruent with the Raup and Sepkoski data set.

The periodicity idea was off and running but there was an even more incredible fact: the periodicity hypothesis was not even original. Seven years before, in 1977, Al Fischer and Mike Arthur (of Bottaccione Gorge palaeomagnetic fame) had published a paper entitled 'Secular variations in the pelagic realm' which had already identified periodicity in the mass extinction of marine faunas. The difference, however, was that the Fischer and Arthur compilation had not had the benefit of the rigorous quantitative analysis that Raup and Sepkoski could bring to bear on Sepkoski's data set.

The debate about the periodicity phenomenon rumbled on through the 1980s more or less acrimoniously. There was one notable occasion when the Royal Society convened a meeting to discuss the state of the art. Star guest Dave Raup was brought in from Chicago to give a talk on the current status of the 26-million-year extinction concept. The room was dead silent as the cycle guru stood to give his peroration – but expectation soon turned to incredulity when Raup started with a very dull collection of musings on the completeness of the stratigraphic record! It was like turning up at Indianapolis for a ride round the track in an eight-litre monster only to be offered a trip round the car park in 'Del-boy' Trotter's three-wheeler instead. By the end of Raup's presentation, the hostility in the room was tangible. Sir Arnold Wolfendale – Astronomer Royal and holder of the most important scientific office in Britain – himself stood and reminded Raup that he had risen from his bed at five that morning to travel to London for this meeting in general, and Raup's presentation in particular, and he for one was not satisfied. He invited Raup to answer the question: what was the current status of the extinction periodicity hypothesis? Raup, taken aback, said merely that he regarded the 26-million-year cycle hypothesis as more compelling than ever. He declined to be more specific. That was the end of reasoned debate at that meeting. Simon Conway-Morris, palaeontologist extraordinaire and Cambridge professorial heir-apparent, read *The Times* with obvious boredom while all around him scientific discourse collapsed. The breakdown was complete when Charles Holland, Head of Geology at Trinity College, Dublin stood up and said that he was very sorry but as far as the impact periodicity hypothesis was concerned he simply didn't believe a word of any of it.

The bullet hole in the world

While the debate about periodicity rumbled on, the big problem of the Berkeley group remained: where was the crater? The chemical composition of the spherules found in the various K–T boundary sites around the world suggested an oceanic origin for the ejecta from the impact. So for most of the 1980s the Berkeley group thought that the impact crater was under-neath the ocean. Since they were looking for a hole at least 10 km across the impact crater must lie in one of the few areas of the ocean which had not been properly explored, for example the Southern Ocean. An alternative theory was that the impact crater had been subducted: consumed at the continental margins. However in 1984, Bruce Bohor, a quietly spoken but tenacious geologist at the US Geological Survey in Denver, Colorado, led a study that found shocked quartz in a number of non-marine K–T boundary sites in the continental United States. The evidence was compelling. Quartz is a continental mineral that is not present in ocean crust so, if the impact had been in the marine realm, where could the quartz have come from? To explain both the marine and the terrestrial evidence by the mid-1980s the idea was mooted that there had been two impacts, one on land and one under the sea. The one under the sea could have been subducted, but there should at least be evidence of the terrestrial crater. This was when attention turned to the Manson impact structure underneath the Iowa farmland. Although an early estimate suggested that the Manson crater might have been about the right age, in the early nineties, detailed radiometric dates on freshly drilled material proved beyond doubt the crater was 74 million years old, predating the K–T boundary by some nine million years.

So by the mid-eighties not only was there conflicting evidence as to whether or not the asteroid impact scenario was real in the first place (the Dartmouth group were being as vociferous as ever), there was also debate as to whether the impact had been in the ocean or on the land! Also by the mid-eighties, dozens if not hundreds of K–T boundary sites had been analysed around the world including two in the general area of the Caribbean: at Beloc in Haiti and in the Brazos River in south Texas. The Beloc section was remote and had not been intensively studied whereas the Brazos River section had been studied for oxygen and carbon isotopic change across the boundary. Unfortunately the K–T isotopes from the Brazos River section showed clear evidence of diagenetic change and were therefore considered to be unreliable. As a result, interest in the Brazos River site waned, especially among the geochemical community. But there was one thing that both sections had in common, and that was massive sandy deposits exactly at the K–T boundary itself. The sandy deposits had been noticed before, but it was the Dutch palaeontologist Jan Smit who finally realised its significance in the Brazos River section. The sandy deposit at Brazos – sandwiched between limestones above and below it which indicated sedimentation in quiet waters – could only be evidence of major disruption to the prevailing sedimentary regime. Jan Smit said that the sand layer at the Brazos boundary had to be evidence of a giant tidal wave or tsunami. South Texas was somewhere close to K–T ground zero.

At about the same time, a Haitian geologist, Florentine Maurasse, reported an unusual rock layer of K–T age outcropping at the southern end of his native island. Maurasse had interpreted these deposits as being of volcanic origin, a description which, given the volcanism/asteroid impact polarisation of the debate, stimulated a young Canadian geologist, Alan Hildebrand, to travel to Haiti to study the deposits for himself. Hildebrand had been intrigued by the K–T mystery for several years and had schooled himself in the various types of K–T layer that had so far been discovered. He recognised instantly that the rock they were looking at was not of volcanic origin at all. Packed with spherules and shocked quartz it could be only one thing: impact ejecta from the asteroid. Using algebra that related the thickness of the ejecta deposit to distance from the impact site Hildebrand was able to calculate that the crater could not be more than a thousand kilometres from Haiti or Texas.

Hildebrand searched for the impact crater using subtle variations in Earth's magnetic field. Initially he was intrigued by a buried structure on the margin of the Colombian basin, but then found something that looked much more likely – a more massive crater further to the north on the Yucatan, half in and half out of the ocean.

The irony is that this strange semi-circular structure on the north-east margin of the Yucatan peninsula had been known about for at least three decades before the Alvarezes came up with the asteroid impact scenario. The Mexican national oil company PEMEX had started searching for oil after the Second World War and by the fifties magnetic and gravity surveys had shown up a buried structure about 180 km across centred near the coastal town of Progreso.

PEMEX thought that the buried structure might be the outline of an oil-bearing reservoir. However, the drill cores hit basement – hard rock unlikely to harbour oil – at about 1500 metres and when samples were brought back to the surface they were classified as the volcanic mineral andesite. To PEMEX executives it looked as if the Yucatan structure was just another extinct buried volcanic caldera. And nothing happened until the late 1970s when an American geophysical consultant, Glen Penfield, was hired by PEMEX to remap the Yucatan peninsula. Flying overhead in his survey plane, Penfield measured very high magnetism and low gravity in the centre of the structure as well as an outer ring of lower magnetism and higher gravitation. The two sets of data clearly confirmed its circular shape. Penfield realised that this looked like an impact structure and as early as 1981, only a year after the Alvarez paper in *Science* that started the whole ball rolling, Penfield with his Mexican collaborator, Antonio Camargo, presented news of the feature at a meeting of the Society of Exploration Geophysics in Los Angeles. At that meeting they even made the connection between their impact feature and the extinction of the dinosaurs! The idea was taken up and championed in the local media but made only a small 'impact' with the scientific community. Walter Alvarez himself was transiently interested but the news that the original cores through the buried crater had been lost in a warehouse fire caused him to lose his enthusiasm very quickly. This was the time when any impact crater was assumed to have been buried at sea. Bruce Bohor's shocked quartz and the implication that the crater might have been on *land* was still some years in the future. The time was not yet right to connect the Yucatan with the death of the dinosaurs.

All that changed in 1990 when Alan Hildebrand, now alerted by the finds in Haiti and connecting their significance with the sandstone beds in the Brazos River, met up with journalist Carlos Byars at a meeting in Houston. Byars told him of the Penfield and Camargo presentation almost a decade before and Hildebrand immediately found and read the original abstract. In growing excitement he sought and met with Penfield. After discussions that had soon expanded to include Penfield's old friend and collaborator Antonio Camargo as well as Mexican research geologist José Grajales, they discovered that some of the original drill cores had not been destroyed in the fire; some were stored at the University of New Orleans and others were back in Mexico in the vaults of IMP – the research arm of Pemex. Using this material, Hildebrand found that a core from *outside* the crater showed clear evidence of shattered rock (known technically as a breccia) as well as shocked quartz, while drill core material from *inside* the perimeter of the crater showed the characteristic minerals expected from rock that had been fused by an asteroid impact.

The implication that the crater – which they named Chicxulub (Chic-shoe-lube) after a local fishing village nearby – was evidence of the K–T 'smoking-gun' was overwhelming. In his original paper in the high-profile American journal *Geology* in 1991 Hildebrand had presented chemical data from the core inside the crater wall that positively identified it as melted rock from an impact but he knew that the ultimate proof would only come from radiometric dating. If the crater and the K–T boundary sections in Haiti and Brazos River turned out to be the same age it would be positive proof linking the two. The study was led by Carl Swisher, an expert in geochronology at the University of California at

Berkeley who had pioneered the technique known as 'single crystal laser fusion', a development of the potassium–argon dating technique discussed in Chapter 3. The advantage of the single crystal laser fusion technique was that it would work on tiny samples – single mineral grains from within the rock matrix. This in turn meant that multiple measurements could be made on a single rock sample and the measurements would not be an amalgamation of data from several different mineral grains. Each date could be scrutinised and evaluated separately.

The team obtained three separate ages which when averaged gave a date for the crater of 64.98 million years. The error was plus or minus 0.05 million years: 50,000 years, which is less than the blink of an eye in geological time. The team then dated spherules from Haiti and found an age of 65.01 million years. The error was a similarly tiny 80,000 years. The two dates were effectively identical. Cause had finally been unambiguously linked with effect and from that moment in 1992 onwards no serious palaeontologist denied that the impact that had made the Chicxulub crater was also the cause of the death of the dinosaurs.

The Terminators

Extinction of an idea

These days the extinction periodicity hypothesis doesn't get much air-time in palaeontological circles. After the initial excitement had faded, cooler contemplation began to show up the huge holes in the whole idea. The periodicity hypothesis stands or falls pretty much on two things: similarity of the causative mechanism at the different extinction horizons – asteroids or comets smacking into the Earth and wiping out life – and regularity of spacing of extinctions in the geological record. But in fact only two occasions have been found in the geological record where there is unequivocal evidence for impacts: at the K–T boundary and again at a mass extinction horizon close to the Eocene–Oligocene boundary (about 35 million years before present). And in the latter case it seems that there were several impacts spread out over more than a million years – not one cataclysmic event analogous to the one that wiped out the dinosaurs.

One of the nails in the periodicity coffin came when people started to examine other, older extinction horizons within the context of the Alvarez–Raup–Sepkoski 'asteroid killer' framework. The problem was the fact that Raup and Sepkoski had relied on one particular timescale – one version of the GPTS – to generate the regular spacing of extinctions (this was the so-called Harland timescale) and in the early eighties timescales were changing faster than house prices. It wasn't long before other timescales had presented revised estimates of the ages of period boundaries (and therefore mass extinction horizons) and, particularly in the early Mesozoic, the idea of the apparently regular spacing of mass extinctions started to fall apart.

The extinction horizon which attracted the most attention was the biggest one of them all – the boundary between the Permian and the Triassic periods (which we shall call the P–Tr boundary). The Permian is the last period of the Palaeozoic era and the Triassic is the first period of the Mesozoic, so the Permo–Triassic boundary, like the Cretaceous–Tertiary (K–T) boundary, is an era boundary – that is, the faunal differences between the two are so extreme that they are used to identify two of the four major chunks of Earth history. The P–Tr extinction dwarfs the K–T extinction. It is estimated that over 50 per cent of all families of organisms alive on Earth became extinct at the P–Tr boundary and over 90 per cent of all species living in the oceans. So why is there all the interest in the K–T extinction if the granddaddy of them all lies ignored? The answer is partly expediency, the K–T boundary is simply more amenable to study because, being younger, there are more K–T boundary sections out there (they have not been degraded by diagenesis or subducted by plate movements). But once the furore over the iridium anomaly had been started by the Berkeley group and then accelerated by the periodicity jamboree of Raup and Sepkoski, people started to look carefully at this more ancient boundary.

Drilling for the end of the world. Gartnerkofel, near Reppwand, Austria, 46.30N, 13.15W. Sometime in the late 1980s

This high up the cold wind blowing round the shoulder of the mountain cut through Bill Holser's fur-lined jacket like a scalpel even in the height of the Austrian summer. To the south the curtain of snow was still blowing off the summit of Mt Sernio. The snow plume seemed to be a permanent feature, or at least it had been in the three days since the scientific party had arrived. It formed a shimmering veil against the pale azure of the

Italian sky. Further to the south the still bluer pool of the Adriatic lay in the crescent arms of Italy and Yugoslavia. The air was crisp, the view spectacular. The mountains of the Carnic Alps towered all around him, a jagged and overlapping sequence that looked like the tank-traps of the Second World War, isolating him from the rest of the world. This was the best part of being a palaeontologist, the places that you visited, the scenery you saw; in short: fieldwork.

But to appreciate this place properly required imagination. You needed to see it with the four-dimensional eye of the palaeontologist. And that was what he could see in his mind's eye now. Before it had been thrust into the sky by the Alpine mountain-building episode, he imagined it when it was still a shallow sea close to a shoreline thrusting out into the western edge of the transglobal superocean called Palaeo-Tethys – the 'father' of the Mesozoic era's Tethys – and therefore the 'grandfather' of the present-day ocean: the Mediterranean. In the days of the Palaeo-Tethys, the continents of the world were not as they are today. If man had been around then it would have been possible for him to walk from the Arctic to the Antarctic – a 12,000-mile journey across the supercontinent known as Pangea – the land mass that had been all the world of the late Permian, the last period of the Palaeozoic era.

And that was what brought him and the team here today. That deceptively simple question: just what was it that made the Permian the last period of the Palaeozoic? The argument was at least vaguely circular, the Permian was the last period of the Palaeozoic because its top was defined by a major mass extinction horizon, one that was big enough in fact to dwarf even the better-known K–T boundary. Two hundred miles due south a straight line drawn from Holser's current position would intersect the south-east-trending coastline of Italy, further on it would bisect the flat plain of the Po valley, further and further over foothills and then the steepening mountainsides of Italy's central spine until it arrived over a small town nestling in the Apennines: Gubbio; and just up the valley from there to the north-east the clay layer in the Bottaccione Gorge, the progenitor of their current enterprise.

That was it, that was why they were here, to see if they could repeat the Alvarez team's success. The sound of a big engine starting split the quiet. The truck that had brought the rig up from Bolzano was reversing into position, grinding backwards until it was perfectly positioned over the white markers painted on the grass, then with a hiss of hydraulics the stabilisers winched down on to the frozen ground, the deck of the truck tilting and settling until it was perfectly flat, the frame of the rig now stark against the sky and surrounding mountains. A concentric circle of scurrying activity as the scientific crew ran around, hauling cables between the rig-truck and the electronics van which housed the gear the geophysics guys would use to monitor the down-hole parameters – principally gamma ray resistivity which would tell them the density of the rock they passed through, like a shorthand signature of the different formations penetrated until they reached the rock layer that was their target. A few miles to the west, in the direction of Tesero, were the local outcrops that had given their name to the formation that they sought far beneath their feet. The Tesero formation was the boundary bed, the thin division that separated the vast thickness of the Permian Belerophon formation from the overlying Triassic Werfen formation. The names of the narrower time divisions that these rocks represented were if anything even more exotic, the Dorashamian, the last period of the Permian, and the Scythian, the first period of the Triassic. All over the southern Alps parts of these formations were preserved but the overall thickness of the sedimentary sequence in this area – the very thing that made it valuable in fact – meant that all that could be seen were bits and pieces, small fragments of the total picture. And yet this succession in the southern Alps was the thickest and most continuous section of this age in all the world. Never mind the rumours that the newly discovered succession in South China was as good; even if that were true the palaeomagnetic reconstructions of the plate positions in the late Permian showed positively that the south China plate could not have collided with ancient Pangea at that time. Therefore its faunas could not be taken as representative. South China had been an island continent – a refuge that had finally collided with Pangea in the early Triassic – it had missed the main action. So, if they wanted to understand the extinction at the Permo–Triassic boundary, then they needed an 'edge', a new angle on an old problem. And, after years of trying to organise financial and logistical support, they had made it. They were finally going to get the 'edge' they needed – they were going to drill right through the Triassic and into the Permian.

If the Permo–Triassic boundary would not come to them, then they would go to it.

The road kill on the mass-extinction superhighway

There were two main reasons why Bill Holser and his colleagues had decided that they needed to drill the P–Tr boundary. The first of these was the incredible scarcity of outcrop sections that spanned that particular boundary. This stems mainly from the fact that the biggest sea-level regression (or fall) in the history of the Phanerozoic coincides with the P–Tr boundary. In fact, for years the prevailing wisdom had been that it was this that was responsible for the mass extinction at the boundary itself. Falling sea level had removed the habitats of the finely adapted shelf-sea faunas and simply killed them off. For the old-school palaeontologist it was an easy solution – you couldn't find the geological sections – and the reason was the same as the thing that had caused all the dying. A perfectly circular argument that was good for nothing as a scientific hypothesis.

The other reason that had stimulated the drilling at Gartnerkofel was that the few good outcrop sections (sections that are found *on* land) through the P–Tr boundary suggested that the detailed time sequence of events was very confused. This may seem a contradiction; how can the biggest mass extinction in the history of life on Earth be 'confused'? The problem was that as the few available sections were investigated and the data compiled it was becoming clear that the P–Tr extinction was not neatly packaged into one media-savvy apocalypse like the K–T boundary extinction. The events that collectively comprised the P–Tr boundary extinction seemed to be messily spread across several million years and different ecological responses in many different groups of animals and plants.

It looked like a sequence of road kills after an Australian road-train run.

The lack of sections and the hints of a complex structure to the extinction event itself made the P–Tr problem rather hard to get a mental hold of. To get around these problems Holser and the others had formed an international team of scientists to drill in the one area known to have the best of a bad lot of outcrop sections. If they were lucky they'd get a good core and solve one of the biggest mysteries of palaeontology.

Why was the P–Tr extinction so important? It is disingenuous to say simply that it was the biggest extinction in the last 570 million years. It does not do justice to the scale of the catastrophe. We need to calibrate ourselves with reference to the K–T boundary and remember that the P–Tr boundary was an *order of magnitude* more severe. The Palaeozoic was the longest of the three eras that make up the Phanerozoic – some 345 million years long. The combined duration of the Mesozoic and the Cenozoic is only 225 million years, 65 million of which belong to the Cenozoic. The remaining 160 million years are the Mesozoic's, the era of middle life. So the long duration of the Palaeozoic tells us that its flora and fauna were basically very successful. The shallow oceans surrounding the continents were host to a great diversity of invertebrates, many of which lived above the sediment–water interface (i.e. they were epifaunal) and extracted nutrition by filtering passing plankton and detritus from water currents. Some of the most spectacular of these were the stalked echinoderms (blastoids and crinoids), distant relatives of our modern-day starfish and sea-urchins. These make up some of the most spectacular of all invertebrate macrofossils and because they could not move around, their bodies could grow to a massive size, developing forms that rival modern-day plants in their extraordinary complexity and beauty. Closer to the shore, giant reefs, built up of slab-like or rough-surfaced corals, dreamed in sunlit waters. The free-swimming citizens of the Palaeozoic seas were shoals of molluscs (orthocones and goniatites), accompanied by the occasional free-swimming trilobite.

In older textbooks of palaeontology the seas of the Palaeozoic are portrayed as tranquil vastnesses where nothing much happened, the childhood of life, easing itself towards the teenage excesses of the Mesozoic and then the adult extravagances of the Cenozoic. By the Devonian, the bony fish had managed to produce a variant that could flop in and out of the muddy waters close to land – the lobe-finned fish. By the end of the Palaeozoic these had evolved into amphibians and beyond, into reptiles and primitive mammal-like reptiles. The arthropods that had started the Palaeozoic with the trilobites and eurypterids had also enjoyed a successful business relationship with the environment and had produced the insects – who had then proceeded to exploit the marketing opportunity of all time – flight.

The flora of the Palaeozoic was long-established and successful too. Land plants had colonised the continents by the late Silurian, scrubby and leafless at first, barely more than green stems scrabbling to fix carbon from the atmosphere and not daring to stray far from water. But by the beginning of the Permian this 'Palaeophytic' flora was complex and diverse. The Carboniferous saw the acme of this floral regime, vast tracts of land supporting rainforests of ferns and primitive seed-plants. Continent-sized forests remained virtually untouched since herbivorous land-dwelling vertebrates were too scarce to do any damage. So let us not underestimate the success of the denizens of the Palaeozoic, they were doing very nicely and, if whatever happened at the P–Tr boundary had not happened then our life at the top of the Holocene would have looked very different indeed.

The Mesozoic started with a few lean millennia as the world came to terms with the fact that something had denuded it of most of its inhabitants, but by the mid-Triassic the oceans were starting once again to teem with life. But now the near-shore sediments were dominated by infaunal feeders (invertebrates that lived below the sediment–water interface), principally molluscs of the bivalve and gastropod type. The epifauna (invertebrates that lived on top of the sediment) of the Palaeozoic seas were gone. The trilobites were gone too, as were the corals that had made up, for example, the beautiful reef complexes of the Wenlock. To be sure, a few things had survived the great dying at the P–Tr boundary, *some* crinoids, *some* bryozoans, *some* free-floating molluscs that would go on to evolve into the ammonoids. But these groups, the masters of the Palaeozoic world, were vastly reduced in diversity. Their dominance was gone and they inhabited the seas of the Mesozoic only on sufferance. On land, the tetrapods and the mammal-like reptiles were gone too, thereby clearing ecological niches that within a few million years would start to be filled by the dinosaurs. The flora too had radically changed, the green plants (pteridophytes, pteridosperms, cordaites) that had made up the forests of the Carboniferous had been replaced by a 'Mesophytic' flora of conifers, cycads and gingkos as well as new groups of pteridophytes and pteri-dosperms.

The old order changeth – but the question is how fast? And that brings us back to our road-kill metaphor and Bill Holser's core-hole in the Carnic Alps.

The core was a spectacular success. It entered the Werfen formation (the Scythian age of the lowermost Triassic) 57 metres below ground level, and penetrated the Belerophon formation (the Dzhufian and Dorashamian ages of the uppermost Permian) at 231 metres. The hole bottomed out when it met more resistant rock at 331 metres. Oolites – tiny spheres of calcium carbonate formed inorganically in shallow water environments which are well mixed by wind and waves – were found at the bottom of the Werfen formation which matched those found in the local outcrop sections near Tesero. One common and age-diagnostic fossil – a conodont rejoicing in the name *Hindeodus parvus* – was found in the bottommost 40 metres of the Scythian sediments recovered from the hole, which suggested that Holser and his colleagues had indeed recovered a very thick sequence through the P–Tr boundary. But the deciding factor in evaluating how useful the core would be came from geochemical techniques. Carbon and oxygen isotopic measurements were the first to be made and it was immediately apparent that the ratio of carbon-13 to carbon-12 showed a truly spectacular decline across the P–Tr boundary. Such a decline had been hinted at by the few measurements that had been made on outcrop sections in the local area but nothing like this had ever been seen before from any P–Tr boundary section in the world.

And there was more, the carbon-13 minimum was accompanied within a few metres by no fewer than two others: there were multiple carbon isotope blips – excursions – spanning the boundary. The oxygen isotope data were more equivocal, reflecting some effect from the dreaded 'big D'. But there was still enough structure in the curve to show a 6 part per thousand enrichment in the light isotope of oxygen – oxygen-16 – at the same time as the lightening of the carbon isotope ratio. Holser and his team made measurements of the ratio of the rare earth elements cerium and lanthanum as well as looking for any enrichment in iridium. The iridium data were a busted flush for the asteroid fans. There were two peaks but they didn't make it above the parts-per-trillion level, a thousand times less than the iridium level found in the K–T boundary sections. But the rare earth data on the other hand *were* interesting.

In the 1980s, the pioneering work of two English palaeontologists, Tony Hallam and Paul Wignall, had shown that sediments laid down in oxygen-poor water had relatively more cerium (Ce) than lanthanum (La). They had developed this ratio into a subtle tool for recognising the presence of anoxia – marine oxygen deficiency that is a known killer in the fossil record – even in the absence of other sedimentary clues. A Ce/La ratio of 0.5 indicated marine conditions that were oxygen-rich (or oxidising); anything over 1.0 suggested oxygen-poor (reducing) conditions. Sure enough, where the Holser team found the oolites the Ce/La ratio was about 0.5 but further up the core, precisely synchronous with the two profound carbon isotope decreases, the Ce/La ratio rose above 1.0. Additionally, these two intervals were accompanied by distinctive layers full of pyrite crystals – 'fool's gold' – a mineral known to form only in conditions of oxygen starvation on the sea floor. So the data suggested that something significant and very, very bad had indeed happened.

Holser and his colleagues put it together into a neat picture. They suggested that the three light-carbon events near the boundary were caused by the overturn of deep waters that flooded on to the continental shelves where the bulk of late Palaeozoic life lived. This water was deficient in the heavy isotope of carbon (carbon-13) because it had been swamped by isotopically light carbon (carbon-12) that had dropped into it through aeons of ocean surface productivity. By the same token the water was also deficient in oxygen – for all the dissolved oxygen had been consumed returning the organic carbon to solution (oxidising it). And so this stagnant water had flooded the shelves – the powerhouse of Palaeozoic life – and killed it. Just below the first carbon isotope blip the ocean had been well-oxygenated and full of vibrant life, the oolites were proof positive of that. And then the poisoning episode began. With an almost eerie beauty the Ce/La data were in full accord. Where the carbon isotopes and the oolites suggested well-oxygenated conditions the Ce/La ratio suggested well-oxygenated conditions. Where the carbon isotopes went light and the oolites disappeared the Ce/La ratio went above 1.0, evidence that the sediments were overlain by stagnant water. Bill Holser and his colleagues had their story – the end-Permian extinction was the result of some kind of overturn in the ocean. But they could not ignore the fact that at least some of the extinctions that comprised the mass killing associated with the P–Tr boundary took place on land. To explain this, they suggested quite reasonably that the deep water that had been overturned had released vast quantities of carbon dioxide (the principal gas of decay in stagnant water bodies) and that this entered the atmosphere inducing a global warming that had killed off the indigenous terrestrial vertebrates.

The Holser scenario was fine as far as it went – but the problem was that it didn't go far enough. It was true that many marine families and more particularly their sub-taxa (orders, genera and species) had become extinct at the P–Tr boundary, but other crucial ingredients that made this extinction the big daddy of extinctions were *not* concentrated near the boundary. First and foremost there was the emerging observation that the tetrapods and the therapsids were not exterminated conveniently near the boundary. The land-dwelling vertebrates – which we can think of as being of advanced amphibian yet retarded reptile persuasion – reduced in diversity in at least four waves of mini-extinctions over most of the duration of the Permian. The final *coup de grâce* to a few of their number occurred in the Dorashamian, the last stage of the Permian, but after three preceding mini-extinction episodes, frankly, so what? Four hard times spread across at least ten million years do not a mass extinction make . . .

On top of this were the plant data. The Palaeophytic flora had been changing towards the Mesophytic flora since the beginning of the Permian when the supercontinent Pangea had begun to coalesce and ice had begun to form at Earth's poles. The rise to dominance of the Mesophytic flora had been similarly gradual. The insects showed signs of a major catastrophe in the lateish Permian but their fossil record is so patchy that it is hard to nail down the timing accurately. What does seem likely, however, is that with the flora in transition the insects too would have had to change, some groups becoming extinct and others radiating to fill vacated ecological niches.

For years the prevailing wisdom had suggested that the cause of the extinctions around the P–Tr boundary had been the grouping together of the continents to form Pangea. As this happened, the amount of available continental-shelf habitat space shrank accordingly and resulted in an increase in the rate of extinction. This is a surface area-to-volume argument; many differently sized, separate islands and continents mean many differently sized and separate fringing continental shelf habitats. So one large land mass will have a smaller habitable shelf area-to-land ratio than many smaller land masses even if the land area remains the same. And in the case of Pangea we must remember that not all of the fringing shoreline would be suitable for supporting a shelf fauna confined to tropical and temperate latitudes. But this explanation does *not* explain the 'pulsed' nature of the several mini-extinctions that collectively comprise the P–Tr crisis, nor does it explain the changes in the terrestrial vertebrate fauna and the indigenous flora unless they too were dependent on climatic changes associated with the formation of the supercontinent.

And so the current state of the P–Tr boundary debate is an unsatisfactory mishmash. A messy road kill of both scientific hypothesis and extinction in the animal and plant world. The best that can be said is that the extinction at the P–Tr boundary seems to be an amalgamation of several different but complementary effects that were spread out over several million years in the late Permian and early Triassic. The faunal change between the two periods – distinct enough anyway to have caused a boundary to be placed there – appears to have its origins in some form of deep ocean disturbance that delivered a worldwide *coup de grâce*.

Watch this space . . .

The day it rained forever

The next time you cross the border between England and Wales you'll probably be travelling on the new suspension bridge across the River Severn. You'll be able to see, looming through the mist to the north, the outline of the old, original Severn Bridge. Spare a thought then for the rock which underlies its easternmost pier. It outcrops in a small cliff at the village of Aust and the rock is filled with pebbles and other rocky debris. Technically it is known as a conglomerate. This conglomeritic rock contains fossils in abundance – particularly of reptiles and primitive mammals – and they are often broken and rarely attached to other bones. These sad, disarticulated ruins were strange enough to attract the attention of Bob Savage who put Joe McQuaker, a graduate student, to work on them in the early eighties. This rock is of Rhaetic age – the uppermost stage of the Triassic period which, as we have seen, started so messily with the road-kill mass extinction. McQuaker and Savage concluded that the assemblage probably represents the wreckage of marine animals which were washed ashore after a severe storm. Above the Rhaetic is the Jurassic, so the top of the Rhaetic is therefore another period boundary and another mass extinction event but, much like the events that surround the P–Tr boundary, it is only the tip of an iceberg of destruction.

The P–Tr boundary is significant for two reasons, first because it is traditionally seen as the biggest mass extinction of the Phanerozoic and secondly because it was the point at which Raup and Sepkoski became sufficiently confident in dating their extinction events to suggest the 26-million-year periodicity hypothesis. The Triassic–Jurassic (Tr–J) boundary is the next mass extinction horizon up towards the Recent and is also a biggie. Like the P–Tr boundary, the sequence of events is confused with several small extinctions seemingly linking together to produce an apparent mass extinction event. However, one of these extinctions stands out, and this occurs in the two rock stages that immediately underlie the Rhaetic: the Carnian and the Norian. Although at the time of their compilation (1984) it was known that there was some kind of extinction event somewhere around this time, Raup and Sepkoski declined to recognise it as a separate entity, preferring instead the notion that the Carnian–Norian extinctions were somehow related to the big events around the Tr–J boundary. With more modern data it is now clear that the Carnian extinctions occurred in two discrete phases, the earlier part affecting the marine faunas, and the other affecting the land-dwelling vertebrates close to the Carnian–Norian boundary. There is something special about these vertebrate extinctions, for they vacated the ecological niches that allowed one particular group of land dwellers – the dinosaurs – to diversify. But how did this happen? Was it some extraterrestrial catastrophe such as marked the K–T boundary? Was it the spread of oxygen-starved waters on to the continental shelves that delivered the *coup de grâce* and altered climate as at the P–Tr boundary? Surely it must be something special – it must be apocalyptic! Something suitable is needed to enter into the era that ended with the crater of doom.

Worlds may not end with a bang but with a whimper – and it seems that worlds need not begin with a bang either, a whimper will again suffice. But the age of dinosaurs was started not so much with a whimper as with a whisper.

The whisper of falling rain.

As we have seen, getting to grips with events in the geological record is the ultimate exercise in circumstantial evidence. Cause and effect are almost never clear-cut and it is terrifyingly easy to mislead yourself. But sometimes strange ideas are the only answer – witness the Alvarez hypothesis – an apparently wacky idea that has stood the test of twenty years of the most antagonistic inquiry (I am reminded here of Sherlock Holmes' famous dictum, 'It is an old maxim of mine that when you have excluded the impossible, whatever remains, however improbable, must be the truth.') And in the late eighties two palaeontologists came up with an idea to explain the late Carnian extinctions on land and in the sea: it rained. Hard and for a long time, to be sure, but basically it rained. They relied on two complementary sets of evidence. The first came from a classical geological discipline, sedimentology: the study of ancient sediments. The other was speleology: the study of caves. Michael Simms (another of Bob Savage's old students) and Alastair Ruffell observed that one of the most distinctive units within the upper Triassic Keuper Marl Formation is a widely developed sandstone body called the Schilfsandstein. The Schilfsandstein is recognised not only in Germany but also in Britain where it is called by various names (one of many charming yet confusing traditions of geology that stems from its diverse nineteenth-century origins as the private preserve of warring gentleman scientists), the most well-known of which is the Butcombe Sandstone Member of the Bristol area. This location is, as we shall see, significant. The Schilfsandstein shows evidence of 'channel bedding' where ancient rivers have cut down into the surrounding sediment soon after it was laid down. The frequency of this channel bedding indicates that there was an increase in the number of rivers feeding the oceans of the Triassic period, at least in the European area. This is particularly noticeable since the Triassic was otherwise a very arid period. The supercontinent Pangea began to break up only in the early Triassic and while this was happening vast areas of the newly forming continents remained to a large extent desert. So any evidence of enhanced rainfall tends to show up like five aces in a poker game. In addition the Schilfsandstein grades laterally into a different type of mud and carbonate-rich rock which has a particular ratio of clay minerals that is known to occur only in wet and humid environments: specifically the proportion of 'kaolinite' increases relative to 'illite'. Kaolinite has a couple of significant economic properties: it is a major component of the clays that go to make fine bone china – the Carboniferous Etruria Marl that underlies the city of Stoke-on-Trent is rich in kaolinite and hence this area has made some of the finest and most desirable bone china in the world. The great potteries of Stoke were built on its availability – Spode, Wedgwood, Doulton, Ainslay made their fortune and generations of antique hunters happy with the kaolinite of the Etruria Marl. Another major use of kaolinite is as a stomach antacid. So next time you overindulge, spare a thought for the clay mineral that ushered in the age of the dinosaurs!

The other piece of evidence that suggested that the Carnian was a time of enhanced humidity and rainfall comes from the age distribution of late Triassic caves and the sediments that fill them. The south and south-west of England are characterised by a great sweep of some of the hardest of the UK's native limestones – the Carboniferous Limestone, or as it is commonly abbreviated, the Carb Lime. This limestone is atypically hard, so hard in fact that it contains stylolites – compression structures where the ancient sediment has not only undergone the normal compression and alteration into limestones but has then been further compressed by either great heat or pressure or both until the original fabric of the rock has been squeezed into thin white lines that look like scar tissue. This rock has been compressed beyond reasonable endurance – it practically cries out its pain. But there is one erosive force that even limestone as hard as this cannot endure, and it is millennia of simple rainfall. Limestone can never win because of pH – acidity and alkalinity. Limestone is alkaline – it has a pH greater than 7. Rainfall is (to a greater or lesser extent) acid because of the carbon dioxide dissolved in it. It is weak carbonic acid. The pH of rainwater varies through all manner of causes. In times of high atmospheric carbon dioxide (the Cretaceous or early Eocene) rainwater will be more acid than during times of lower CO_2, for example the late Ordovician. But the point is that rainwater is always more acid than limestone and under certain conditions – when marine limestones are uplifted forming terrestrial terrains for example – they can form cave systems as the rainwater worries away at cracks and holes in the fabric of the rock and gradually converts them into subterranean streams and rivers. Such a terrain is called 'karstic' and much of Croatia is made up of just such landscape. The Carb Lime outcrops of southern England are also karst country. These ancient fissures are ideal for the speleologist (cave explorer), and Bob Savage was for many years president of the University of Bristol's speleological society. But there is a problem with dating karst deposits, as any surviving fossils in the limestones are likely to be so distorted as to be unusable, and even if they were of use they would only tell you when the original sediment was laid down, not when it was converted to limestone or when it was karstified.

However, there is one method that provides a lower limit on the age of karstification. It relies on the fact that these fissures are almost always filled with sediment and often these are fossil-bearing. Simms and Ruffel examined many of these so-called fissure fillings themselves and also checked the literature. They found that the sediments which infilled the karstic caves were all Norian – the next stage up from the Carnian – or at most early Jurassic. The conclusion was inescapable: any reasonable estimate of the amount of time needed to fill in these cracks in the ground was at most a few million years, a duration that on the timescale of palaeontology is virtually instantaneous. So they concluded that sometime in the Carnian it had rained hard and long enough to induce cave formation in the limestones of southern Britain and that then it had simply stopped, allowing sediment to infill the caves rather than being washed away.

So it appears that events occurring at the end of the Triassic were, much like those around the Permo–Triassic boundary, a mishmash through which discrete events are only now beginning to emerge. In the case of the P–Tr boundary the identifiable killing mechanisms are volcanism (the eruption of the Siberian Traps), and the shoaling of oxygen-deficient water on to the continental shelves. In the case of the Tr–J boundary, it appears that this 'pluvial' (rainy) episode caused a major change in the prevailing climatic regime to which the existing flora and fauna were ill-adapted and so paved the way for the dinosaurs.

The black thumb-prints of Earth's strangler

One of the most important of Raup and Sepkoski's extinction events was in more or less the middle of the Cretaceous, 90 million years before present. The horizon is well-exposed in Britain and is worth a visit if you are ever in either the Dover or Humberside areas. In Shakespeare Cliff at Folkestone there is a more or less monotonous cliff of chalk interrupted about halfway up by an enigmatic band of flint. In this Chalk succession the shells of the forams are in relatively poor shape – they are infilled by calcite that has been dissolved, transported in ground water and then re-precipitated inside the shells forming a type of contamination that is known to the trade as 'spar'. Close examination of the biostratigraphy (the succession of fossils) at Shakespeare Cliff suggests that there may be missing sediment between the upper reaches of the Plenus Marls (the rock unit at the bottom of the cliff) and the Melbourne Rock (the unit at the top).

There is a gap in time.

The sequence at the Humberside locality, most famously exposed in South Ferriby pit almost in the shadow of the Humber Bridge is superficially similar to the Folkestone section in that it is in a Chalk succession. However, halfway up the white cliff face the sequence is dramatically interrupted by a band of almost pure black about a metre thick. This is known, somewhat unimaginatively as 'the Black Band'. The normal succession of nannofossils, foraminifera and invertebrate macro-fossils is interrupted here and instead all that is to be found are remnants of organic rich shale. How can we explain this? The standard explanation for black shales in the fossil record is anoxia. In the absence of oxygen (that is, reducing conditions) this unrecycled organic detritus begins to accumulate and decay into a stagnant mess on the ocean floor.

It was in the 1970s that two sedimentologists – Seymour Schlanger of Northwestern University in Illinois and Hugh Jenkyns of the University of Oxford – began to compile data about the timing and geographical occurrence of these black shale outcrops in the Mesozoic. They realised that the Cretaceous seemed to have been especially prone to intervals of oceanic stagnation. The picture that was emerging was pretty terrifying. They identified and numbered no fewer than three intervals of the Cretaceous when the entire ocean apparently became stagnant. They called these episodes Oceanic Anoxic Events (OAEs). The first of these, OAE1, was a lengthy interval of several million years in the early Cretaceous which spanned the Aptian and Albian stages. The sediments of the area south of Grenoble in southern France are the remnants of this stagnant ocean. In an area known as the Vocontian Trough, which stretches between Vaucluse and Digne in the Alpes de Haute Provence, vast tracks of dark brown to black shale form scrubby foothills. It is a terrain that is in some ways reminiscent of Wyoming. In various places the foothills of monotonous shale are interrupted by beautiful beds where narrow limestone ribs only a few centimetres thick alternate with black shales. The regularity of these rock couplets is astounding and is now known to reflect the same Milankovitch cycles that have controlled the succession of glaciations in the past few million years (discussed in Chapter 4). In other words the same orbital rhythms which dominate our planet's climate today also ruled the world of the early Cretaceous, a world without ice. The limestone ribs reflect a time when the deposition of anoxic sediments was interrupted by depositional conditions with more oxygen.

Schlanger and Jenkyns also identified a minor, short-lived event in a late Cretaceous stage known as the Santonian that they called OAE3. But what of OAE2, the event between the Aptian–Albian and the Santonian events? OAE2 was the jewel in the crown, the oceanic anoxic event to end them all. OAE2 occurred at the Cenomanian–Turonian (C–T) boundary and the Black Band at South Ferriby is its remnant. OAE2 is better documented than the other OAEs. It shows its sedimentary spoor not only in South Ferriby but also in the Umbrian Apennines. Far below the K–T boundary, halfway down the hill in the direction of the deserted bar is a metre-thick outcrop of black shale. It is known in Italy as the Livello Bonerelli – the Bonerelli horizon – and biostratigraphic and radiometric dating shows that it is precisely contemporaneous with the Black Band at South Ferriby. And there is more: at DSDP Hole 585 in the Pacific Ocean a narrow band of black shale is also to be found a quarter of a kilometre below the sea bed and once again it is precisely at the Cenomanian–Turonian boundary. The black thumb-print of OAE2 is global.

What happened in the mid-Cretaceous?

In order to grasp the causes of the Black Band we have to understand a little more about carbon isotopes in fossil (and living) organisms. As we have discussed previously, the factor that partitions carbon isotopes differently into organic material and inorganic material is *photosynthesis*. But beyond that, the factors that control the distribution of isotopically light and heavy carbon are more varied. In the ocean, the light isotope of carbon (carbon-12) is removed from surface waters when organic material dies and falls towards the sea bed. This light carbon is then usually returned to the ocean's pool of dissolved carbon in deeper waters. Palaeontologists refer to this recycling of light carbon within one reservoir (the ocean) as 'within-reservoir variability' and when averaged over the whole ocean the proportion of the two carbon isotopes does not change. But the proportion of light to heavy carbon in the ocean *can* be changed by external factors – something that is known as 'between-reservoir variability'. The most common between-reservoir change is when organic carbon (carbon with a disproportionately high level of carbon-12) is buried in sediments. This leaves ocean waters enriched in the heavy isotope of carbon, carbon-13, resulting in an unusually positive carbon isotope composition in the animals and sediments formed in these waters. When the chalky remains of these animals or sediments are measured sequentially through time this 'heavy' composition shows up as a positive kick in the carbon isotope curve.

It is possible to dissect the nature of these events in greater detail by analysing benthic and planktonic forams in a way that is precisely analogous to the approach used in the Pleistocene. When this is done for the Cenomanian–Turonian boundary – and there are relatively few places where it can be done successfully given the great antiquity of the sediments – benthic and planktonic forams appear to become enriched in carbon-13 at approximately the same rate and by the same amount. The fact that the planktonic signal parallels the benthic signal so precisely tends to suggest that surface ocean waters were not acting independently of deeper ocean waters, and that the predominating effect was the burial of organic carbon and its removal from the ocean. We are forced to conclude that, at the Cenomanian–Turonian boundary, conditions conducive to the burial of organic carbon became favourable, perhaps because of a marked decrease in the rate of deep ocean ventilation that prevented the organic carbon from being oxidised, or perhaps because surface ocean productivity increased so much that more organic material was produced than could be oxidised. In either event, the deep ocean's supply of life-giving oxygen was used up and the deep waters of the world's oceans stagnated.

The detailed reasons for such widespread anoxia at the Cenomanian–Turonian boundary remain an enigma but there is good evidence to suggest that the events around this time had profound and long-lasting consequences for climatic change over the next 90 million years.

As figure 7.1 shows, the record of global temperatures from the oxygen isotopes in the limestones of the Bottaccione Gorge rise gradually from the early Cretaceous towards the Cenomanian–Turonian interval, peak at the boundary and then start a downward trend that continues until the present day.

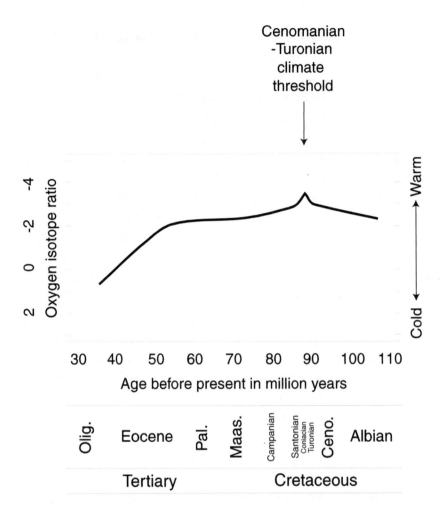

Fig. 7.1. The temperature record of the late Cretaceous and early Tertiary reconstructed from the oxygen isotope record from the limestones of the Bottaccione Gorge. Note the critical threshold at the Cenomanian–Turonian boundary.

At the start of the twenty-first century we still don't know what really represents an 'anomalous' climate. Parts of the past 570 million years – the interval that we have reasonably good fossil evidence for – appear to have been marked by intervals of climatic warmth – the Cretaceous is a good example and perhaps part of the Palaeozoic too. And yet what we fashionably call the 'ice-house' climate of the late Cenozoic is not alone in the fossil record – there is good evidence for glaciation at the end of the Ordovician, for example. However the data that we have at present makes it clear that the Cenomanian–Turonian boundary represents the most important climatic threshold of the past 100 million years. Something happened then, and it is disingenuous to suspect that it was unrelated to the carbon burial episode. The best guess currently is that the black shale at the boundary – the Black Band in South Ferriby pit, the Livello Bonerelli in the Umbrian Apennines, the nameless dark horizon in DSDP 585 and all the other exposures of the C–T black shale around the world, represent a time when organic carbon was buried, and with it another form of carbon – carbon dioxide. The removal of this greenhouse gas from the atmosphere would have led rapidly to a decrease in temperature and would have preconditioned the world to climatic cooling.

The Cenomanian–Turonian boundary is not alone in the geological record in having this type of effect. Two more recent episodes appear to reflect the same thing, although one is very much more complex than the other.

The big chill and the prelude to man: the middle Miocene

To take the simpler of these two first we must move far uptime to the middle Miocene, only 14 or so million years ago. The middle Miocene is characterised by a carbon isotope change to *heavy* values which also seems to have had a profound influence on our present-day climate. Indeed it is no exaggeration to say that if the C–T event started preconditioning Earth for the big chill it was the middle Miocene event that pressed the 'go' button. The middle Miocene carbon 'blip' is neither as large nor as rapidly occurring as the C–T boundary. It is not even as large as the enigmatic Palaeocene event. But it is still significant and its structure is similar to the C–T event in that planktonic and benthic fossils responded similarly. So it too was a carbon burial event and therefore it too sucked carbon dioxide out of the atmosphere and consequently cooled Earth. And this precipitated some drastic climatic consequences. At this time Antarctica had finally separated from Australia and South America and had become thermally insulated from them by a ring of waters known as the circum-polar current. As Antarctica drifted across the south pole it entered one of the two coldest areas of Earth. It was primed for ice formation. When middle Miocene carbon burial started removing Earth's insulating blanket of CO_2, Antarctica started cooling and, since it was not receiving heat from any other nearby continents, it just kept on cooling. Gradually snow began to fall and that snow turned to ice. The reflectivity of Antarctica increased turning away more solar radiation and so the continent cooled still further until eventually two ice sheets formed: the tiny ice sheet on Antarctica's western arm and the enormous East Antarctic ice sheet on the rest of the continent. And this was the true beginning of the late Cenozoic age of glaciation, for with that fortuitous clumping of ice the world eased much closer to the knife edge that separates the icehouse world from the greenhouse world. The record of Earth's jolt towards cooler temperatures is plain to see in the oxygen isotope record of middle Miocene benthic foraminifera – a sudden cooling of about 4°C (the exact figure is complicated by the uncertainties imparted by the ice-volume effect discussed in Chapter 4). It is the biggest single deep-water cooling event of the past 30 million years, a final move towards the icehouse Earth. When ice started to form in the northern hemisphere 2.5 million years ago, the grip of the icehouse world intensified. Since the late Pliocene we have been completely locked into the glacial–interglacial regime, the record of the oxygen isotope stages (pioneered by Emiliani and Shackleton) which reflect these glaciations now extends well back into the Pliocene.

And what of the middle Miocene event? Is there some sedimentary record of this event analogous to the Black Band at the Cenomanian–Turonian boundary? Yes, there is, it extends all the way around the Pacific basin and is precisely contemporaneous with the middle Miocene carbon isotope excursion and the sudden cooling seen in the benthic foram oxygen isotope record. It outcrops too on the Big Sur coast, not far from where Jack Kerouac decided to hit the road. It is called the Monterey Shale.

The outpost of the Cretaceous

We return now to deeper time, to the Palaeocene, the first period of the Cenozoic, which is, as it happens, well-represented by rock outcrops in the badlands of Wyoming. The Palaeocene was a time of recovery. The dinosaurs

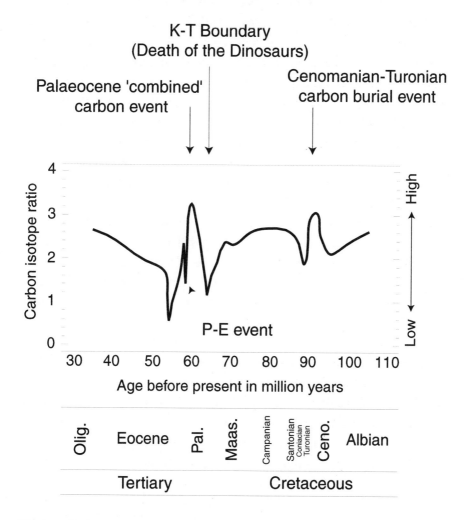

Fig. 7.2. Carbon isotope record from the limestones of the Bottaccione Gorge for the late Cretaceous and early Tertiary interval. The carbon isotope maximum during Palaeocene time (as well as the sharp minimum associated with the Palaeocene–Eocene boundary event) is clearly seen.

and many other animals both on land and in the sea had become extinct at the K–T boundary and so something else had to take their place. On land this was the mammals who had already been around for a considerable chunk of the reign of the dinosaurs but had not been able to do much to improve their lot with all the niches that they could exploit already occupied by reptiles. In the oceans the Cretaceous plankton, both nannofossils and planktonic forams, had almost all become extinct while enigmatically the *benthic* forams had escaped apparently unscathed. During the Palaeocene the chemistry of the oceans was also out of kilter as a result of the big kill at the K–T boundary, and nowhere is this better observed than in the isotopes of carbon. The carbon-13 record of the Palaeocene is singular: it increases from the minimum of the K–T boundary to levels that look very like those of the Cretaceous and then, in the early Eocene, decreases to very negative values that for the rest of the Cenozoic never again even approach those of the Cretaceous. What can this mean?

In detail, using the approach of analysing both planktonics and benthics we see that the planktonic record increases more substantially than the benthic record. This increase in the *difference* between the surface and the deeper water records shows that the Palaeocene excursion was at least partly a productivity event. But the benthic record increases too which suggests that the Palaeocene excursion was *also* a carbon burial event. This in itself is a striking difference from the C–T and middle Miocene events. The oxygen isotope record of the Palaeocene shows evidence of profound climatic changes, particularly in the higher southern latitudes. As carbon-13 ratios increased during the early and middle Palaeocene the temperature of surface waters in the high southern latitudes decreased. At the same time, we have evidence for a cooling of only *deeper* waters in several other areas of the world's oceans. Conversely, as carbon-13 ratios decreased from the late Palaeocene to the early Eocene high-latitude surface waters warmed. Global deep waters followed suit. The link here between carbon isotopes and climatic change is the same as for the C–T and middle Miocene events. As productivity and the rate of carbon burial increased during the early to late Palaeocene both marine and atmospheric carbon dioxide levels decreased. Temperatures therefore

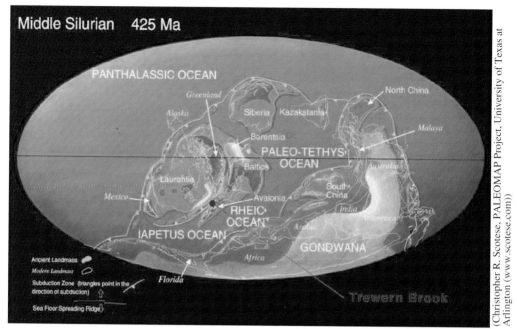

Geographic reconstruction of the world during the Middle Silurian. The position of the sediment that would one day be the section at Trewern Brook is shown

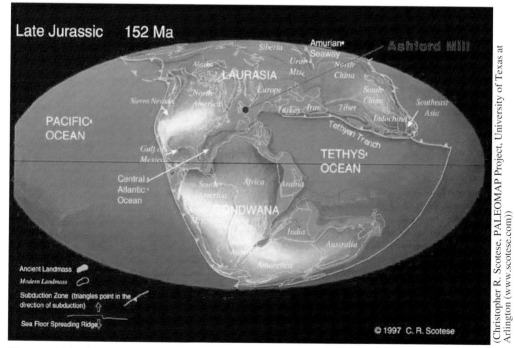

The world in the late Jurassic. The position of Ashford Mill is shown

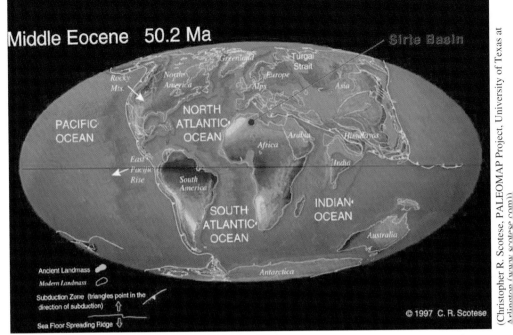

The world in the Middle Eocene showing the position of the Sirte basin. Arabia and northern India are ocean and North and South America have not yet collided. The resulting circum-equatorial ocean is Tethys

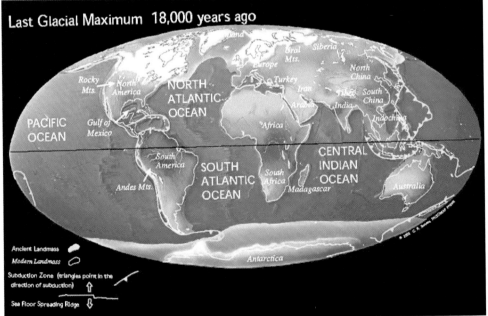

The world during the last glacial maximum. The huge ice sheets have depleted the oceans in the light isotope of oxygen – the resulting signal will be measured by the Chicago mafia

(Michael Durkin)

Shadwell Quarry in the Welsh borderland. These rocks clearly show the boundary between the Wenlock (the grey limescale below) and the Ludlow (the brown shales above) series

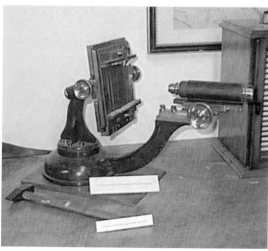

(Richard Corfield)

Lapworth's microscope in the archives of the University of Birmingham. It was especially designed for the examination of graptolites

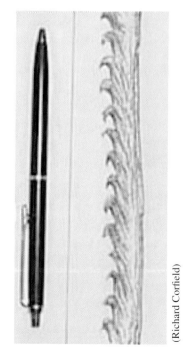

(Richard Corfield)

One of Ethel Wood's original drawings of a graptolite

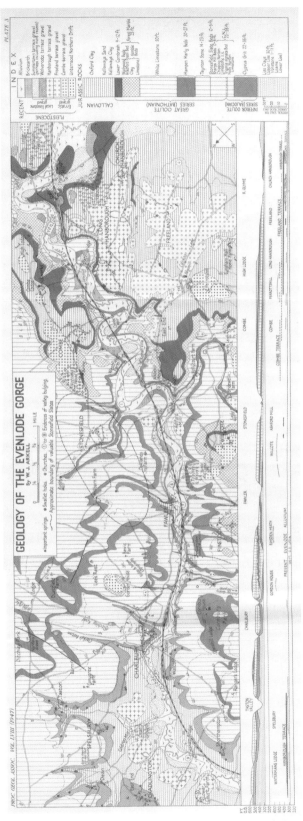

Arkell's map of the Evenlode Gorge in Oxfordshire. The quarry at
Ashford Mill is near the tiny village of East End. Arkell's use of
colour shows that the older rocks outcrop at the north-west with
progressively younger rocks cutting the surface to the south-east

The 'flight-tube' of Aston's mass spectrograph

A light microscope image of a nannofossil. A member of the group called discoasters

The cover of *Science* announcing the discovery of the 'Pacemaker of the Ice Ages'

(Richard Corfield)

The K-T boundary at Stevn's Klint

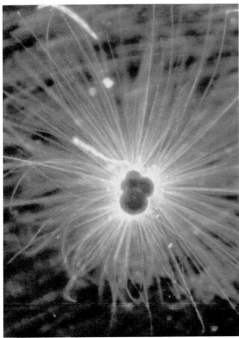

(Howie Spero)

Living planktonic foram

(Richard Corfield)

The author at the K-T boundary layer in the Bottaccione Gorge

(Ocean Drilling Program)

The Ocean Drilling Program's ship: JOIDES *Resolution*

(Richard Corfield)

The chalk face at South Ferriby pit. The 'black band', a relic of mid-Cretaceous ocean stagnation, is clearly visible

Emile Zuckerkandyl

Svante Pääbo

(University of California at Berkeley)

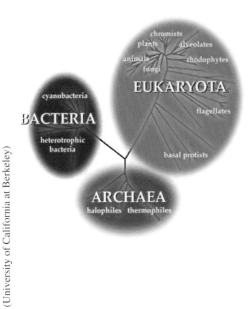

Carl Woese's Universal Tree of Life
showing the three domains

Cesare Emiliani

(University of Miami Rosenstiel School Archives)

decreased, starting in the high latitudes. Conversely, during the late Palaeocene, as productivity and the rate of organic carbon burial decreased, so carbon dioxide levels in the ocean and in the atmosphere as well as high latitude and deep-water temperatures started to increase again. There appears to have been a degree of overshoot associated with this increase and the early Eocene is known to be one of the warmest intervals of the Cenozoic as a result.

But what of this influence on deep-water temperatures far from the high southern latitudes? It seems that the agency for this control was heat transport by deep waters. Deep waters had begun to form in these latitudes by the middle of the Palaeocene if not earlier. As they sank and flowed away northwards, insulated by their vast mass, they carried with them the temperature of their formation. The thermal changes occurring in the high latitudes were thus transmitted to the deep waters of the rest of the world by messengers from the Antarctic.

The details of Palaeocene climatic change are only now beginning to be unravelled. They remain a tantalising enigma. But there is one small segment of the transition from the Palaeocene to the Eocene that has attracted even more attention.

Academic hunger and deep ocean indigestion

The eighties drew to a close and the nineties dawned. Ronald Reagan staggered out of the White House, out of public recognition and into presidential extinction. In Britain the prospect of new evolutionary innovation in the political sphere grew as it became clear that Mrs Thatcher's days were numbered. It was time for a change in the realm of palaeontological extinction theory, too. Iridium anomalies – even the big kill at the Permo–Triassic boundary seemed suddenly passé. It was time for something new. And it came from the deep south, from Maud Rise in the Southern Ocean, a few hundred miles north of the Ross ice shelf in Antarctica. The occasion was Leg 113 of the Ocean Drilling Program and on board were two ambitious palaeontologists. At that time Lowell Stott was a graduate student of Jim Kennett, a guru of palaeoceanography. All on board were primed for some big discovery. This, after all, was one of the first legs in the history of deep-sea drilling that had been aimed specifically at the high latitudes, which had been left virgin by the predecessor of the Ocean Drilling Program (the Deep Sea Drilling Project, or DSDP) because the DSDP's drilling ship, the *Glomar Challenger*, had been unable to drill in polar waters. The *Challenger* was the sister ship to one with the ultimate chequered past. The *Glomar Explorer* had been built for the US Government to retrieve a Soviet nuclear submarine that had gone down in deep water in the Pacific Ocean and had been paid for by none other than Howard Hughes (he of the multi-millions and the chicken soup fixation). The *Challenger* though, eventually became the flagship for a new enterprise in geology – it became the drill ship that recovered the cores for the Deep Sea Drilling Project. But eventually the *Challenger* had been retired and its place was taken by a custom-built drill ship hired from the oil support giant Schlumberger – the Sedco BP 471 – colloquially known as the *Resolution*. The *Resolution* started work on Leg 100, the first leg of the ODP, but Leg 113 was the first to venture into really high-latitude waters. The scientific team on Leg 113 were determined that they were going to find something that would change the world.

They were indeed primed.

Two holes were drilled on the Maud Rise in the Southern Ocean, only 300 miles from the Ross ice shelf, one of the two main regions where deep waters are formed today. Like all DSDP and ODP holes they were given numbers and it was ODP 689 and ODP 690 that together would change the world. Both had good K–T boundary sections and Stott and Kennett, after analysing this remote sequence, published their findings in *Nature*. But the K–T boundary had been done to death by then (Alan Hildebrand was already on the verge of rediscovering the Chicxulub crater), and something else was needed to re-ignite interest in the new science of fossils, and it was to be oxygen and carbon isotope analyses at the next period boundary up from the K–T, the Palaeocene-Eocene (P–E) boundary. For years, this had been considered a period boundary of apparently stupendous boredom dropped into the middle of a sub-era of the Cenozoic known as the Palaeogene. The first period of the Palaeogene (and therefore the Cenozoic) is the Palaeocene and the second is the Eocene. The similarity of their names is a warning about the troublesome nature of these two time intervals, for the Palaeocene was only split off from the Eocene rather late in the history of stratigraphy, in the early twentieth century. Despite its exotic carbon-induced climate changes the Palaeocene is the orphan period of the Cenozoic – unwanted on the one hand by its predecessor the Cretaceous which finished, as we have learned, with a hell of a bang – and its successor, the Eocene, which has its own uniqueness as a period of anomalous warmth. For years the Palaeocene was ignored by the great and the good as that bit of the Cenozoic that happened between the action of the end-Cretaceous and the serious business of the Cenozoic, which was when global cooling got going in earnest, ultimately culminating in the late Cenozoic glacial ages and the evolution of a group of tiresomely quarrelsome simians.

Since the Palaeocene wasn't of much interest, the boundary between it and its big sister the Eocene got pretty much ignored until 1990. But, as always in the business of fossils, the dictates of fashion and the sudden availability of new material meant that it was just about to be propelled into superstardom. Stott and Kennett were the agents who sold the world the P–E boundary.

Jim Kennett had just set up an isotope lab at the University of California at Santa Barbara where he was now Dean of the Faculty of Geosciences. Isotope labs always need a dedicated staff, although it is fashionable to pretend that the instruments are so automated that they run themselves, and Kennett had a young graduate student (a scenario that may sound familiar) who helped set up the lab for him, Lowell Stott. After they returned from Leg 113 they analysed the samples through the K–T section and up into the Cenozoic from ODP Holes 689 and 690. The sediments at ODP 689 were deposited in deeper water than those of ODP 690 and this difference gave the game away.

Sea temperatures should reasonably be expected to decrease with depth even in the Arctic and the Antarctic where deep waters are currently formed and the vertical temperature difference between surface and deeper waters is at a minimum. Kennett and Stott found that benthic temperatures from these two sites – identical in every way apart from the depth at which the sediments were deposited – were, at the P–E boundary, *reversed*. Deep waters were warmer than shallower waters. (This approach is a very minor modification on the isotopic technique for tracing the pathways of fossil deep waters discussed in Chapter 4.)

Kennett and Stott speculated that there were effectively three types of ocean. The first of these was the thermohaline ocean that we're familiar with today where deep waters are formed in the high latitudes by cooling and sinking. They named this type of ocean 'Oceanus'. The second type of ocean was the so-called halothermal ocean, where (*see* Chapter 4) deep waters are formed in the low latitudes by salinity-induced sinking. They named this ocean 'Proteus'. They also envisaged a third type of ocean, a kind of halfway house where both modes of deep-water circulation might hold sway that they called 'Proto-Oceanus'. Their idea was that the thermal inversion at the Maud Rise sites was caused by an atavistic episode in the Proto-Oceanus ocean of the early Cenozoic when the newly predominant mode of thermohaline circulation changed back to the old system of halothermal circulation and surface waters in the low latitudes were, for a limited time, once again the major supplier of deep waters to the world ocean. Coming as it did so soon after the K–T debate, Kennett and Stott were not backward in imagining cataclysmic consequences for their deep ocean circulation change and cast about for any evidence of mass extinction in the deep ocean at that time. Their search, however, was disappointing. The only paper that then existed with any relevance to the events at the P–E boundary was a minor work by Tjalsma and Lohmann (the latter of later eigenshape analysis fame, *see* Chapter 5) concerning an apparent decrease in the number of species of benthic foraminifera more or less at the Palaeocene–Eocene boundary. This, although of minor academic interest, was hardly likely to set the world alight. Fortunately, there was someone else on board the *Resolution* that Austral summer who had some relevant information. Ellen Thomas was the benthic foraminiferal micropalaeontologist on Leg 113 and once she heard about Kennett and Stott's story she examined the benthic foram fauna with redoubled vigour across the well-preserved boundary interval in the two Maud Rise holes. She found that there was a dramatic decrease in diversity of the benthics right at the boundary, thereby refining and enhancing the Tjalsma and Lohmann story.

The original papers by Kennett and Stott were published under various permutations of authorship priority in the new version of the 'blue book', the *Scientific Proceedings of the Ocean Drilling Program*.

By now Kennett and Stott had even found a pronounced but dramatically short-lived negative carbon isotope excursion at the P–E boundary which was not dissimilar in magnitude (although very much shorter in duration) to the excursion at the K–T boundary. What is more, the oxygen isotope record showed a major negative inflection at exactly the same moment. The stage was set – they had two excursions and a story – and they were ready to make their splash. However, even with its new look, the 'blue book' was still not quite the same as having an idea published in a high-profile, rigorously peer-reviewed journal. So Kennett and Stott submitted their paper to *Nature* in 1991. It was accepted, and the P–E bandwagon started to roll.

The story was quickly embellished when Jim Zachos and Paul Koch touched base with the Mecca of palaeontology and found the same carbon isotope excursion in the Palaeocene–Eocene deposits of the Big Horn Basin of northwestern Wyoming. This was important because it was in a completely different depositional environment from the deep ocean which in turn carried the clear implication that what Kennett and Stott had found was some form – as they had suspected – of global signal in the world's budget carbon. The impetus for Zachos and Koch's visit to the Big Horn basin came from the fact that both were based at the University of Michigan as post-docs (scientists who have just completed their doctoral studies) in the early nineties and that Michigan used Wyoming as a field area for training undergraduate students. Zachos and Koch went out there to help as summer field assistants and realised – as Marsh and Cope had before them – that the wonderful diversity of Wyoming's rocks and fossils could help them with their own studies. A literature check and several conversations with vertebrate palaeontologist Phil Gingerich confirmed that there was indeed a P–E boundary section out there in the big country and so they went and sampled it. The material that they measured was the fossil remains of teeth from the tiny proto-horses that had roamed the Wyoming plains in the 20 million years following the K–T boundary as well as carbonate soil nodules. They presented their findings at the 1991 Geological Society of America meeting in San Diego. Jim Zachos decided to present the carbon isotope data leaving Paul Koch the oxygen isotope data. The latter were far more equivocal than the former, in fact the record was so noisy that it was simply not possible to be sure that a real change equivalent to the Kennett and Stott oxygen isotope shift existed at all. The overwhelming likelihood was that the material had been so altered by lying in Wyoming's scrubby foothills for 50 million years that it was all diagenetic overprinting (the big 'D' discussed in Chapter 4). The carbon isotope data though were unequivocal: there was a major negative excursion right at the position of the P–E boundary (as documented using the local biostratigraphy of Gingerich and his colleagues). Like the K–T boundary beforehand the pace of investigation speeded up as more and more investigators piled on the bandwagon and started measuring isotopic change across the P–E boundary. In Oxford, I pulled our carbon isotope profile from the Bottaccione Gorge out of its bottom drawer and a took a close look at the P–E boundary interval. Sure enough there was one data point that was much more negative than its neighbours: we'd had it all along and never attached any significance to it, assuming wrongly that it was just a part of the natural 'noise' in the data set. But Kennett and Stott had done more than just note a change at the P–E boundary. They had made measurements on both planktonic and benthic foraminifera and found that the carbon isotope signal in both had changed by equal amounts.

As we have already seen, the surface and the deep ocean are normally decoupled and therefore have different carbon isotope signatures which reflect the differing influences of productivity and deep-water ageing at the surface and at depth. If both records did the same thing . . . exactly what did it imply? A young geochemist called Jerry Dickens was the first to come up with an explanation, and Jerry's big idea hit him in a bar.

Michigan's Department of Geology has taken the bar/pub tradition of palaeontology to new heights. Rarely a day passes that faculty and students don't troop across the road to the bar after work, in a way that is strangely reminiscent of the American servicemen's legendary interest in afternoon beer call. This turns out to be no coincidence when you discover that some of Michigan's most glittering faculty have backgrounds in the US military. There is something distinctly wholesome and no-nonsense about Michigan's attitude; if you're at Michigan you live, breathe and work geology and palaeontology. The atmosphere is more that of a family than a university department and this is why Michigan is a hothouse of new and radical ideas and why the post-docs who go to work there love it and don't want to leave. Jerry Dickens was one such post-doc and it was during afternoon beer call, while discussing just what the hell could have shifted both the planktonic and benthic records so far and fast, that an answer dawned.

To understand how he arrived at this solution we need to return to our discussion of 'between-reservoir' effects on carbon isotopes. The most common between-reservoir effect is organic carbon burial that steals light atoms of carbon away from the ocean resulting in a positive carbon isotope excursion recorded in the carbonates crystallised in that ocean. Another class of between-reservoir effects is where carbon is *added* to the ocean, most commonly via river run-off from the continents which puts light carbon derived from land-plant decay back into the ocean, or volcanism which releases light carbon in the form of isotopically negative carbon dioxide from deep within the Earth. Dickens realised that the parallelism of the planktonic and benthic records could only be explained by input of carbon from an external source. Others had suggested that this might be volcanism, but the problem with this was that the magnitude of the change in the carbon isotope records was simply too great to have come from volcanic CO_2.

Volcanic CO_2 has a carbon isotope ratio of about -7 parts per thousand. Given the size of the ocean's carbon reservoir, and the amount of CO_2 that the volcanoes active at the time could realistically deliver, there was no way that volcanism could account for the magnitude of the carbon isotope excursion. (In fact, Dickens' calculations indicated that there was no realistic combination of *any* volcanism that could account for the magnitude of the shift at the P–E boundary.) Dickens realised that there was another source of isotopically negative carbon with a ratio of about −60 parts per thousand and that there was also easily enough of it to account for the excursion. That material is methane hydrate.

Methane hydrates are found on the continental shelves. They are molecules of methane that have been locked up inside a 'cage' of frozen water so that they are trapped. Methane hydrates are very common in the ocean: it is estimated that over a trillion tons of carbon are buried as methane hydrate. The amount of methane hydrate off the coast of Florida and Georgia alone is enough to satisfy the energy needs of the United States for the next 200 years.

Methane hydrates are held in their caged state precariously. They are stable solids only within a very narrow range of temperature and pressure. An increase in temperature of only 5°C or a decrease in the overlying pressure of only a couple of pounds per square inch is enough to release them, allowing them to spontaneously dissociate back into water and methane, and once the carbon in the methane is released it will combine with atmospheric oxygen (within ten years) to form carbon dioxide. When methane is synthesised by natural processes of decay it incorporates vastly more of the light isotope of carbon which accounts for its isotopic composition of −60 parts per thousand, sixty times more negative than the average isotope ratio of carbon found dissolved in the oceans. Jerry Dickens suggested that even a small portion of the world's methane hydrate reserves spontaneously breaking down could easily account for the speed and magnitude of the carbon isotope blip at the P–E boundary. Since the oceans mix on a timescale of only a thousand years or so, the signal would be effectively identical in surface and deep waters, and hence in planktonic and benthic foraminifera. But the huge advantage of the methane hydrate hypothesis was that it explained the negative shift in oxygen isotope ratios as well. In terms of its insulating properties, methane is an even more potent greenhouse gas than CO_2 and what is more its major by-product is carbon dioxide; double jeopardy. Jerry Dickens' story explained both the size of the dual (benthic and planktonic) carbon isotope shift *and* the 8° warming that Kennett and Stott had measured in the high latitudes.

The Dickens story was as big as the asteroid killer scenario had been a decade earlier, and it had more immediate human relevance. For, as more and more P–E boundary sections were found, all showing the carbon isotope excursion, it became apparent that the timescale of this event – an event that had a terrestrial, not extraterrestrial cause – was very short indeed, perhaps even approaching human timescales. The P–E boundary story became another media sensation – a fossil analogue of our own greenhouse future had been found.

By this time, Lowell Stott had moved on to other intellectual territories, disillusioned perhaps by the way that others were crowding in on his discovery. If so, this would have been understandably hurtful, but sadly is typical of all branches of science in our ultra-competitive age. In a way it is also the sincerest form of flattery. But with limited material – always limited with DSDP and ODP cores – together with so many isotope palaeontology labs around the world, the competition for material is intense and ambitious palaeontologists, like Marsh and Cope before them, have never been respecters of priority. Yet it was Lowell Stott and Jim Kennett who discovered the significance of the P–E boundary and their claim to be architects of eternity is therefore assured.

Today the P–E boundary is the most intensively studied interval of the Cenozoic. In 1997, a leg of the Ocean Drilling Program, specifically formulated to investigate the P–E boundary problem drilled the Blake Nose not far from the Georgia coast in the USA and recovered a thick and apparently complete section – ODP 1051 – across the boundary interval that allowed really detailed comparison with the record from 690 which started the whole thing off. My own lab in Oxford has contributed to the story by looking closely at the record, using samples spaced only two centimetres apart. Santo Bains, Richard Norris and myself have found that events at the boundary were not a simple single release of methane. Rather it appears that the methane release occurred in three stages. Our work has also allowed us to shed some light on the causes of the original methane release. The question about what caused the methane to be liberated in the first place boils down to two possibilities: either there was a warming or there was a decrease in overlying pressure either of which took the hydrate outside its stability zone and allowed it to change back into gas. Dickens favoured the former scenario, suggesting that deep waters were generally warming in the late Palaeocene and that eventually they warmed a hydrate reservoir beyond its thermal threshold. To address this Bains and co-workers looked in detail at the ODP 1051 oxygen isotope record for the 300,000 years prior to the boundary and found no evidence for a gradual warming. This led Bains *et al* to suggest instead that a change in pressure must have been the root cause of the catastrophe, perhaps a submarine earthquake had disturbed the overlying sediments, suddenly allowing the buried methane hydrates to return to their gaseous form. Thereafter, warming may have played a role. As methane was liberated into the atmosphere the world began to warm – particularly in the high latitudes where deep waters had already been forming for the preceding 5 million years. These gradually warming deep waters would have found other methane hydrate deposits in the same or perhaps different locations and warmed them beyond their safe threshold. They too would have changed to gaseous methane and carbon dioxide, resulting in even further warming of deep water. The world would have entered a positive feedback cycle of progressive and apparently unstoppable warming. Bains *et al* have suggested that there were three discrete intervals when hydrate reservoirs collapsed. But what stopped the process? What broke this positive feedback loop? They speculate that the answer may lie in a sudden increase in the rate of carbon fixation in the surface waters of the ocean. This heightened ocean productivity would consume the excess carbon dioxide (the end-product of the methane release) very quickly. This would slow, then halt the greenhouse warming and allow the world to return to normal.

How fast did this happen? The answer is very quickly; both excursions are infinitesimal blips in the overall climatic history of the Cenozoic, so quick in the grand scheme of things that they were overlooked until Stott and Kennett found them in the anomalously thick sediment sequences at ODP Sites 689 and 690. Getting accurate dates and durations for the components of these excursions is difficult, for on timescales as fine as this the conventional means of dating sediment samples simply won't work. Normally sediments are dated by using the age estimate of a first or last occurrence of a few well known microfossils – planktonic foraminifera or nannofossils – and then calculating the unknown sediments' ages. But in this case the action takes place so quickly that the nearest microfossil datums are too far away in depth (and therefore in time) in the sediment cores to be of much use. After all, the ages of the microfossil datums are themselves only estimates based on the geomagnetic polarity timescale which is itself more or less loosely pegged to absolute radiometric ages. The conventional means of dating sediments is too coarse. It is not equal to this task.

There are two approaches to date events within this narrow time sequence. The first is to use the theoretical amount of time needed for the ocean to pull the excess CO_2 out of the atmosphere. This so called 'efolding time' approach yields a duration of about 200,000 years for the whole event, from the time of the initial oxygen and carbon isotope excursions to the time that the carbon isotope record flattens off. The problem with this approach is that it assumes that there have been no major changes in surface ocean productivity across this interval, and, as we have seen, this may not be so. The second approach is to take the reasoning of the Pleistocene and apply it to the P–E boundary and use the effectively metronomic variations in the Earth's orbital geometry to constrain dates and durations. An apparently unbroken sequence of precessional cycles has been identified in the P–E boundary sequence at ODP Site 1051 by Dick Norris and Ursula Rohl and since the duration of a precessional cycle is only 21,000 years it has enabled them to develop an independent calibration of time across the interval. Hearteningly, their estimate for the duration of events from the initial excursion to where the carbon isotope trail fades away is also about 200,000 years.

This may sound a lot to us humans, used as we are to dealing in tens and hundreds of years, but to a palaeontologist, it is less than the blink of an eye. And also let us not forget that the onset of the event occupied a tiny fraction of those 200,000 years. In fact if we use that 200,000-year duration to scale the timing of the initial excursion we find that it occurred in less than a thousand years, perhaps even as little as one *hundred* years. The ebb and flow of deep time, through the techniques of the new science of fossils, has finally arrived on the timescales of human comprehension.

And this should give us pause for grave reflection, for the total *warming* at the P–E boundary occurred over perhaps 20,000 years and was 8°C. The initial warming step over the first thousand years was perhaps only a couple of degrees Celsius, but then we need to remind ourselves that the total temperature *difference* between the present interglacial and the last glacial maximum 18,000 years ago was only 4°C. And during the height of that glaciation the ice reached as far south as the Canadian border and covered most of northern Europe. Small temperature differences equal severe and global climatic effects.

The warning from the P–E boundary is stark.

A final question remains: where were the methane hydrate reservoirs that let go all those millennia ago and started the ball rolling? Perhaps it was one of those off the eastern seaboard of the United States close to where ODP Hole 1051 was drilled. That area has its own chequered history; for it is of course the Bermuda Triangle. The enduring myth of the Bermuda Triangle is already interwoven into the fabric of the P–E boundary story. Those with active imaginations have suggested that the several unexplained ship and plane disappearances in the area may have been caused by release of methane gas from buried methane hydrates. As methane is lighter than air, an immediate effect of a release would be to decrease the density of both the water and air overlying the liberated pocket. Ships – and aeroplanes – would sink and drop like stones. Fanciful?

Perhaps . . .

Beyond climate's dead zone

This quick canter through the huge field of extinctions is necessarily incomplete. There is enough material for a book in its own right, and I am woefully aware that I have not covered some of my favourites: the big kill at the end of the Ordovician, the terminator in the late Devonian, to name but a couple.

As for asteroids, there is now good evidence – based on huge tumbled blocks of rock entombed in cliffs in the Nevada desert – for an impact way back in the middle Devonian. Its discoverer, John Warme from the Colorado School of Mines calls this, with a fine sense of irony, 'the Alamo event'. Birger Schmitz, Sweden's most brilliant practitioner of the new science of fossils, has recently re-ignited the impact debate by showing incontrovertibly that in the early Palaeozoic the frequency with which meteorites hit Earth was a hundred times higher than today. Clearly there is still much to be learned from this new field of 'astropalaeontology'.

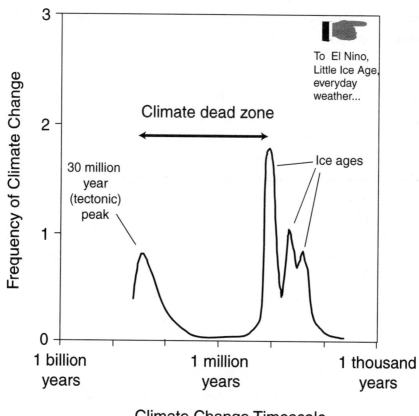

Fig. 7.3. Beyond climate's dead zone. Climatic change can be seen here as occurring on essentially two timescales. One at 100,000 years and less, the other at about 30 million years (more or less the same timescale as the extinction periodicity of Raup and Sepkoski). The large area between the two peaks where no climatic change occurs is the enigmatic 'climatic dead zone'. Note that the bottom axis increases in orders of magnitude.

But before we leave the subject of extinctions, it is worth asking where now is that most radical of all extinction-related ideas, the 'periodicity of extinction' concepts of Fischer and Arthur and Raup and Sepkoski? Jack died tragically in 1999 but the idea that he so enthusiastically promulgated refuses to lie down. It has been recently resurrected by that most indomitable of climate research duos, Nick Shackleton and John Imbrie, who gave it a new spin. By the early nineties the number of long oxygen isotope records that had been generated by the various stable isotope labs working mainly on DSDP and ODP material was nothing less than enormous – automated preparation systems and mass spectrometers had seen to that. The data base was so extensive that Shackleton – the most rigorous of scientists – felt able to process some of these very long records through the spectral analysis software that had originally been developed by Imbrie and his colleagues. Why did he do this? One suspects that Nick – with his long experience of analysing time-sequence records – somehow had a gut feeling that there was something there that needed explaining. And he was right. It was almost eerie. At the highest frequency end of the spectrum, he found the scurrying cycles of climate that contribute to our daily lives, the El Niños, the little ice ages, whose record is left not only in the dead remains of forams but is also to be found in tree rings and ice cores. Beyond that, he found the signature of Earth's journey around the sun, the signature that governs the movements of the huge grinding ice sheets of the glacial ages, the Milankovitch cycles, eccentricity, obliquity, precession, and after that: nothing. No climate variability. A dead zone in the climate frequency spectrum.

But what's that? Out at about 30 million years – almost exactly the same frequency that Fischer and Arthur found, as well as Rampino and Stothers' recomputation of the Raup and Sepkoski data set. Faint to be sure – you can't after all squeeze many 30-million-year cycles into the 100-million-year record of climate change that Shackleton regards as the limit of good data – but there all the same, a climate cycle with a period of 30 million years.

John Imbrie is now retired after a distinguished career at Brown University in Rhode Island and the concept of the 30-million-year climate cycle really represents his swan song. Shackleton and Imbrie declined to speculate too wildly on the cause of their underlying climate cycle, contenting themselves merely with the suggestion that it might have its root cause in some underlying cyclicity of continental drift and landmass reorganisation.

It is of course appropriate that John Imbrie, the man who played such a major role in developing our understanding of climate change in the recent geological past and on high-frequency timescales, should also have a hand in the discovery of climate change on much longer timescales.

And so we leave these two chapters with the feeling that there *is* something out there at about the 30-million-year timescale. A similarity in the frequency of extinction, climate change and cratering ages that demands explanation. What the explanation will finally be only time will tell.

The truth, as the saying goes, is out there . . .

CHAPTER EIGHT

Beyond the Kingdom of the Small

The Gee-Whiz Lab

The USA's love affair with science and technology began at the turn of the twentieth century: Americans proceeded to embrace both with a speed that would have put an anorexic piranha to shame. Specialised institutes sprang up all over the country, most of them funded by extraordinarily generous gifts of private money. Some of these new institutes we have had occasion to mention already: the Woods Hole Oceanographic Institution and the Scripps Institution of Oceanography, for example. Every major centre of learning in the US seemed to want a specialised subsidiary institute devoted to some aspect of the sciences. Most of these focused on understood areas of physics, chemistry or biology, but one of these was unusual. At the turn of the twentieth century when science was not yet seen as inherently multidisciplinary, the millionaire steel baron Andrew Carnegie caught the science bug and decided that an institute that specialised in the gaps between conventional sciences was what was required for the future prosperity of the Union.

The Carnegie Institute was established in 1904 by Act of Congress with an initial capital injection by Carnegie of 10 million dollars (folding money by anybody's standards, especially then!) and was followed by further six-figure donations from the science-captivated Carnegie. Of the fourteen departments established, the first reflected his particular love: an observatory was established at Mt Wilson on the slopes above Los Angeles to further the discipline of astronomy. The other departments were concerned with many aspects of plant physiology, animal embryology, terrestrial magnetism and geophysics. It is geophysics that concerns us, though, and its spectacular successes are owed mainly to one man.

The Paper Polymath. Carnegie Institution, Washington DC, 38.55N, 77.00W. 12 August 1953

He stood in the outer office. He'd been there for about fifteen minutes and that was unusual. The director didn't usually keep his people waiting – especially not the Chairman of the Department of Terrestrial Magnetism (DTM). He eased the crease of his pants and checked his loafers. Polished to a blinding black and white contrast by Neva only that morning, they had not suffered from the hurried walk over from the magnetism building at the director's summons. The girl behind the desk worked quietly through a mound of grey files. She stopped occasionally to glance up at him and smile.

Philip Abelson ran through the reasons why he might be called for in this manner. Was Bush dissatisfied with his work? If so, no hint had been forthcoming. And he had been in the job for almost eight years now. After his work for the navy had finished they had seemed happy enough. A method for the formulation of uranium hexafluoride (which he had patented) had been followed by the chairmanship of the committee on the feasibility of atomic submarines. That had been a success, too. They were already thinking of building them. He shook his head and stared out at the bright spring sunlight. The bustle of cars was audible, rising up and through the open casement from the street below.

The Bakelite phone on her desk trilled. She spoke quietly for several moments, nodded and replaced it quietly, 'Director Bush will see you now.' He was shown into the inner office. The director, a big man with the ready smile of the career administrator, strode forward to greet him. 'Phil. Good of you to come. Won't you sit down?'

Abelson exchanged greetings in kind, thinking furiously and trying not to let it show. This ready greeting didn't suggest any kind of problem. Vannevar Bush eased himself behind the mahogany desk and spread his big hands flat on its empty surface.

'Phil. You know what Carnegie stands for?' It was a rhetorical question, oblique and even more incomprehensible than his welcome. Abelson said nothing. Long years in the Office of Naval Research had taught him the value of zero disclosure.

'It stands for excellence.' The director continued after a moment. 'When Andrew Carnegie endowed this institution it started with fourteen departments. That was whittled down to ten before the Great War and now we're down to five.' He paused significantly, his big hands still on the desk. 'And those five are the best in the world, Phil. Absolutely the best. We have the observatories although Hale tells me that if the light in the night sky of southern California keeps growing they'll have to move to a better dark-sky environment – they're talking about South America. Then there are the biological departments including Cold Spring Harbor, mustn't forget that, and then of course there is the jewel in our crown – the Department of Terrestrial Magnetism. Of which you are the chairman.'

Abelson turned his gaze from the window. 'What about the Geophysical Lab? You didn't mention that.'

Bush's surprise could almost have passed for genuine. 'Ah, yes, silly of me. The Gee Whiz Lab. I'd forgotten that.' Then, 'Bit of a problem there, of course.'

Abelson nodded silently to himself. The Geophysical Lab was a department in need of a destiny, for what his opinion was worth. The other departments had produced some spectacular successes which somehow the Geophysical Lab had never emulated. Indeed, that department had been the one Carnegie had been least keen to fund. He wondered what this could possibly have to do with him.

'I want you to take it over.'

The question, unasked, had been answered. He could not keep the astonishment from his voice. 'Me?'

Bush smiled. 'Why not?'

'I'm a physicist!'

Bush shook his head. 'You're more than that, Phil. I need someone who can delve into new corners, a new broom for the Gee Whiz Lab. I want you to take it over.'

He couldn't believe it. 'But I don't know anything about geology.'

Bush was shaking his head. 'Exactly. That's the point. I don't need someone who knows about geology, I need someone who knows about science. I need someone who can take the GL into new territory. Somebody who can make of it the success that the other departments have been. I'm offering you the directorship of the department. You can take it anywhere you want. And the salary will be commensurate. What do you say?'

A thought flashed across Abelson's mind. It was too soon to call it an idea. It was a nervous spasm. An intellectual vapour. But it just might be worth pursuing. Something that would tie together some left-field ideas that had been knocking around while he paper pushed in the DTM.

'Well?' Bush was shrewd. He could see the wheels turning.

'There is something. But I'll need time to think it over.'

Bush stood, his broad smile back. 'Sure, Phil. Think it over. Take all the time you want. I'll see you back here Monday.'

The man who made a science

Philip Abelson took the job, becoming Director of the Geophysical Laboratory in 1953. He applied himself to the task with the same combination of assiduousness and flexibility that had marked his considerable contributions to science before then. For Abelson had contributed not only to the Department of Terrestrial Magnetism's research in the run up to the Second World War, he had also been the co-discoverer of the ninety-third element on the periodic table – neptunium. During the war he had been affiliated to the Office of Naval Research and had discovered a method of extracting uranium that had been both patented and taken up by the Manhattan Project. Towards the end of the war he had been commissioned by the US government to look into the feasibility of constructing nuclear submarines. His report – finding in favour of the idea

– is largely responsible for the fact that America now has the most comprehensive nuclear submersible navy in the world. Abelson was a polymath like Harold Urey, able to turn his hand to a variety of science problems which crossed discipline boundaries. He took up the challenge that Bush had laid down for him and succeeded magnificently.

In his latter years in the Department of Terrestrial Magnetism, Abelson had set up a group dedicated to the study of biophysics and had endured much ridicule from more narrow-minded colleagues over his ideas that the biology of organisms would somehow be directly related to the laws of the physical universe, an idea that today we find obvious and which is the foundation of Rudwick's palaeontological subdiscipline of functional morphology (*see* Chapter 5). This work may have led him into some conflict with Bush – although from the tenor of the reports from both men it is clear that Bush (at least in public) wanted Abelson in the Geophysical Laboratory on the basis of his unarguable administrative ability.

Abelson took with him concepts that he had been working on in biophysics, particularly those associated with the chemical identification of elements and compounds. He rightly understood that 'geophysics' in the rather strange atmosphere of the Carnegie meant anything old as long as it worked (with the further codicil that it be exciting and epochal) and so decided that he would turn his biochemical and biophysical expertise towards identifying organic compounds in fossil material.

This was an idea beyond radical. There was so little information about the preservation potential of amino acids and other complex organic molecules in fossil materials that people had not even got around to worrying about diagenesis – the dreaded 'D' word – that was already beginning to influence the new fields of stable isotopes in fossil material.

Abelson decided that he would put his DTM techniques for the analysis of modern molecules at the vanguard of his venture into palaeontology. First among these was paper chromatography. Paper chromatography was the earliest technique to be developed for separating molecules. The speed of diffusion of molecules through a porous medium is proportional to the size of the molecule. Hence, if a mixture containing several different compounds is dropped on to a piece of filter paper dampened with a suitable solvent, after several hours the different molecules will have been separated by their different rates of diffusion up the filter paper. If the various molecules have previously been labelled – by dyeing them with different stains, for example, or by using radioactive tracers (another technique that Abelson pioneered) – the different molecules are easily detectable.

Since Abelson was interested first and foremost in testing his technique, he needed a fossil that was both common and would provide sufficient material for him to analyse with the crude techniques available to him. It was known that molluscs secreted their calcite shells from a protein layer. This was an ideal source of organic material since amino acids are the basic building blocks of protein. In order to analyse these very small amounts of protein he would need to start with a vastly larger amount of calcite. Consequently, he focused on large molluscs: oysters and clams.

By the middle fifties, he had successfully showed that a significant proportion of amino acids survived unaltered even in material of Devonian age: 350 million years old. Although this was met with considerable scepticism at first, Abelson's results were unarguable. He showed that even fickle amino acids could survive large tracts of geological time, and laid the foundations for palaeobiochemistry.

Second level in the evolution game

Abelson began to build a research group. One of the first to be attracted to this new area was a young man named Ed Hare who came to work at the Gee-Whiz Lab for his PhD. He was interested in using Abelson's discovery of the longevity of amino acids to address the species problem. This problem arises because the technical definition of a species (as originated by Ernst Mayr) is that all members of the population be able to interbreed, that is, be able to produce offspring and perpetuate the race. Thus it is impossible to recognise a species in the fossil record, because by definition fossils are dead and unable to interbreed. But there is another way of looking at the problem. The physical manifestation of the genotype is the phenotype. And since the genotype – and the perpetuation of it – is what a species is all about, the closer we can get to characterising the genotype, the closer we are to defining a species. If we can do this for several or many species then we would ultimately end up with a taxonomy that accurately reflects true evolutionary relationships. The human genome project can be considered as the beginning of this endeavour, although its fundamental focus at the moment is rightly medical. But that is merely the beginning. To erect a true genetic taxonomy of life on our planet we would need to sequence the DNA of every species on Earth, a number that is estimated to be in the tens of millions. We are used to thinking of DNA sequencing as a new and cutting-edge science, but the truth is that as early as 1970 Ed Hare and his colleague Kenneth King conceived a related approach, based on Abelson's discovery that complex molecules could survive unaltered in the fossil record.

King and Hare formed an effective partnership. King at the Lamont Docherty Geological Observatory, a division of Columbia University perched on the high cliffs by the Columbia River, and Hare at the Gee-Whiz Lab in DC. Their idea was that there was an intermediate way of measuring species differences, something more flexible than merely looking at gross morphology yet at the same time something do-able, not years in the future like sequencing DNA itself. They would look at the intermediate product between the genotype and the phenotype; they would look at proteins. Proteins are the end-product of two processes called transcription and translation that occur within a cell. Transcription is the process by which the DNA makes an intermediate molecule called RNA and translation is the process by which RNA makes proteins by assembling amino acids into long chains called polypeptides which are themselves then assembled into proteins. All enzymes and hormones are proteins and they organise the construction of the body. The sequence of amino acids that governs the structure of a particular protein strand is unique. And this is what King and Hare measured using Abelson's technique. Unlike Abelson (who had worked on molluscs), they decided to use the planktonic foraminifera as their experimental vehicle, for there was abundant material available from the Deep Sea Drilling Project. They focused on core top samples: recent material, and they found, as Abelson had before them while working with molluscs, that there were significant differences in the amino-acid composition between the organisms that they analysed. But King and Hare took it a stage further. They fed their data into a primitive computer – an ancient IBM – and used statistics to see if there was any underlying unity in the data. They found that the amino acids of different species clustered together in a way that exactly matched the existing classification of the same planktonic forams.

Armed with this finding they were then able to apply the technique to fossil foraminifera, analysing the amino-acid compositions of several Miocene species that have no living relatives. Sure enough, after processing the data, they found that their amino-acid signatures matched the existing classification based on gross morphology almost perfectly. It seemed that the world was their oyster and that the technique could easily be applied to other organisms. But King and Hare never took that opportunity.

Just why King and Hare did not take their study further is not clear. Was it because of the large number of forams that had to be picked for analysis? Or was it, perhaps more likely, because bigger and better projects beckoned? Hare went on to do a considerable amount of work on amino acids in higher organisms – including ancient humans from archaeological sites, and developed his lab techniques still further. Their last paper together was in a volume of conference proceedings in the late 1970s entitled *The Biogeochemistry of Amino Acids*.

But the legacy of King and Hare persists. The technology that they pioneered was the first glimmer of a much bigger enterprise that would eventually attract the attention of Michael Crichton and Steven Spielberg two decades later. But first the new science of molecular palaeontology needed to come of age.

Paper-jams in the photocopier of life

The Gee-Whiz group were not the only ones in the fifties thinking deep thoughts about proteins and their constituent amino acids. In California was a man who could claim to be the spiritual father of proteins and their ultimate progenitor, the molecule that stores the genetic code itself: DNA.

His name was Linus Pauling. The story of the two young Cambridge biochemists, Jim Watson and Francis Crick, who discovered the structure of DNA between 1951 and 1953 has become part of modern mythology. But if you have read Watson's tremendous book *The Double Helix* (a true twentieth-century classic, especially in the frank way it treats the ambitions and aspirations that motivate hungry young scientists) then you may recall that the pair of them spent a lot of time trying to find out what Linus Pauling at Cal Tech was doing by pumping his young nephew Peter Pauling for information during his periodic visits to Cambridge. After Watson and Crick published their paper in *Nature* and lined themselves up for the Nobel prize you might have thought that Pauling would have changed direction and gone into other fields, understandably annoyed perhaps at having lost the greatest discovery since *Origin of Species* to two post-docs at the eleventh hour. Nothing could be further from the truth. With hindsight it is possible to argue that Pauling's next endeavours were equally as important. Pauling was not a man to let the grass grow underfoot and he continued thinking about the nature of DNA and how it effected its control on the phenotype. Pretty soon he and others began to understand that DNA functions by copying itself on to a transmission molecule known as messenger RNA (mRNA). Like DNA, mRNA is made up of a sequence of four bases or nucleotides: adenine, thymine (also now known as uracil), guanine and cytosine. There are only two possible combinations of these molecules in the double helix strand of DNA: A-T and C-G. These tiny molecules lock together to form a very strong structural unit held together by a sugar and phosphate backbone. Once the single helix of RNA has been constructed, the exposed, reactive sites on the nucleotides become the template for the assembly of amino acids, the basic building blocks of polypeptides, themselves intermediate structural units between amino acids and proteins. Proteins are a crucially important part of the machinery of the cell because many of them act as catalysts that mediate and accelerate the chemical reactions that fuel life. Without them, life would be impossible as the chemical machinery on which life depends would not proceed fast enough to sustain itself. Such proteins are called *enzymes*.

Watson and Crick, together with Maurice Wilkins of London

University, received the Nobel prize for their discovery in 1962. By a curious coincidence this was the very same year that Linus Pauling and his young post-doctoral collaborator Emile Zuckerkandl founded an entirely new discipline based on their profound understanding of how enzymes are assembled in cells. There is a certain irony here: Watson and Crick receiving a Nobel for work that they had done ten years before; Pauling and Zuckerkandl taking this work further and effectively founding a new science that unified biology and palaeontology.

Zuckerkandl and Pauling pointed out that proteins like RNA and DNA were 'semantophoretic' – a term they coined that translates literally as 'the transmission of meaning'. They argued these molecules are capable of 'talking about their history'. There are three categories of these semantophoretic molecules (or semantides). The primary semantide is nothing less than DNA itself, the secondary semantide is the messenger molecule RNA (in its various forms), and the tertiary semantide is the protein that the DNA-RNA system manufactures. And finally there is the product of these semantides – the organism itself, be it bacterium or elephant – the phenotype. Pauling and Zuckerkandl's radical concept was nothing less than to use the semantides themselves to unravel the history of life itself. They would make these molecules talk.

The theoretical basis for their idea was simple. Pauling and Zuckerkandl knew that some of these large and functionally important protein molecules existed as slightly different varieties which differed to a very small degree in the arrangement of the amino acids that comprised the polypeptides that themselves made up the protein. Functionally, however, these variants were identical. There was no difference in the efficacy with which they carried out their tasks within the cell's machinery. Crucially, the overall structure of these molecules – the most diverse of which were the cytochromes (whose twenty different variants are central to the regulation of cellular respiration) and the haemoglobins and myoglobins (whose half-dozen variants are intimately involved in oxygen transport within blood and muscles respectively) – was so similar that they could not be anything other than homologous, i.e. despite their slight structural differences they were descended from a common ancestor. Since the type and sequence of these amino acids are controlled by triplets of nucleotide bases acting in concert (i.e. genes), Pauling and Zuckerkandl suggested that by mapping the variations in the arrangement of amino acids in these proteins in different organisms they would effectively be measuring the number of mutations that separated the different protein variants. In short, they could measure the evolutionary distance between those species possessing the cytochromes or the iron-containing globins. The molecules themselves preserved a record of their long evolutionary history and by unravelling the sequence of amino acids along their length could be made to tell it: hence the term 'semantophoretic molecules'. In species which possessed both of these protein systems, Pauling and Zuckerkandl would have an in-built double-check of their conclusions.

Pauling and Zuckerkandl published their epochal paper in perhaps the strangest journal in the biosciences: *The Journal of Theoretical Biology. J. Theor. Biol.* remains my personal favourite to this day. I first came across it as an undergraduate at the University of Bristol in the early eighties while working on a dissertation about Lamarckism (the inheritance of acquired characteristics). It was only held in the Medical School library. In a little-used and dusty corner two shelves of the *J. Theor. Biol.* brimmed with strange and outré ideas about the functioning of the natural world. The basic premise of the journal was nothing if not *avant garde*. Here was a journal that thumbed its nose at the usual descriptive and experimental approach of the biological sciences. Here was a journal that clearly saw itself as something special; nothing less than biology's challenge to the physicists with their grand and rather snooty concepts about the physical underpinnings of the universe. So they could have a subdiscipline called 'theoretical' physics could they? Well, heck, biology could have 'theoretical' biology. Why not?

J. Theor. Biol. was established in the early fifties in response to the seismic shift that occurred in the life sciences following the discovery of DNA. The knowledge that biology was entering a new and strictly quantitative world of exotic chemicals whose precise mathematical structure had itself precise and calculable effects on protein manufacture was the spur that biology needed to boot-strap its way on to a par with physics and chemistry. Consequently it could only be a matter of time before the *J. Theor. Biol.* made its own major contribution to palaeontology. It was a mere eight years between the journal's foundation and Zuckerkandl and Pauling's ground-breaking paper.

Not only could the scheme suggested by Zuckerkandl and Pauling address the question of the mutational distance between groups (albeit at second hand, remember, as Zuckerkandl and Pauling themselves acknowledged, the technology to address the coding machinery of the cell itself – DNA and RNA – was not yet to hand) and be used to construct a diagram of probable relationships between these groups without the use of fossils (a phylogenetic tree), it could also suggest the original sequence of amino acids in the ancestral protein that lay at the root of the tree, and most spectacularly of all it could also suggest the *time* at which this ancestral protein existed. The new science of molecular palaeontology could in theory offer not only better information about the relationships between species than classical palaeontology, it could also offer something that had always been the latter's unique advantage: time itself.

The basis for this extraordinary claim was simple. The ten years since the discovery of the structure of the DNA molecule had been ample for biologists to contemplate its potential consequences. One of these was that the structure and function of the molecule simply lent itself to spontaneous change (i.e. mutation). It was inferred that the sequence of nucleotide bases of every DNA strand would undergo regular and spontaneous mutation whereby a base or set of bases would be either deleted, added or replicated. This spontaneous mutation would occur as a function of the activity of the DNA molecule itself, which was forever splitting apart and making copies of itself which then fed the RNA engine of the cell that churned out the metabolically active protein molecules. The variations in the accumulation rate of these spontaneous mutations would, from a statistical point of view (i.e., if observed over a sufficient span of time) iron themselves out, and therefore the number of mutations (as seen in the changed order of amino acids in a protein) that had accumulated would in effect be a molecular chronometer.

To use a metaphor that would not have been available to

Zuckerkandl and Pauling – working as they did in the days before modern office technology – the DNA-RNA complex can be considered the ultimate Xerox machine of life and, like a continuously operating Xerox machine, occasional glitches occur. All of us have had the experience of a sheet jamming for a second in the sheet-feeder and then being fed on to a photocopier's platter at an angle, and subsequent copies – until we stop the machine – will all reflect the irregular alignment of the original. We might stretch the point a bit and consider mutations as the minor paper jams – not enough to stop the machine but enough to produce unusual copies – in the photocopier of life. After a mutation has been introduced in the nucleotide base sequence of the DNA strand, it is then faithfully copied up the semantophoretic chain to RNA and ultimately proteins; two other Xerox machines making faithful copies of the altered original. As we have noted many, indeed most, of these mutations have no effect on the functioning of the protein, consequently there are many different types of the same protein, all of which retain their functional integrity (a pheno-menon known as protein polymorphism). The fact that many different versions of, say, the cytochromes and globins still work is technically referred to as 'degeneracy'.

The mirror crack'd

It was young Zuckerkandl who pursued this new philosopher's stone. He sequenced – using the same chromatographic techniques that had been pioneered by Abelson – the amino acids that made up the polypeptide chains of the haemoglobins and myoglobins in some of the commonest animals known to man – including man himself. The haemoglobin/ myoglobin family is based around an atom of iron. It is the atom that binds to oxygen and confers its unique functionality on this essential respiratory molecule. Each haemoglobin or myoglobin molecule is made up of four polypeptide chains, each itself made up of a string of amino acids. More correctly, these are amino-acid 'residues' since they bind together with the loss of a molecule of common water. And each of these amino acids is coded for by a triplet of nucleotide bases. Potentially Zuckerkandl had the genetic code in the palm of his hand yet it was not the *true* genetic code. It was the genetic code as portrayed through the refractory prism of its expression in a metabolically functional molecule. This point is important because there is unavoidable condensation of information in the transition between the nucleotide sequence of the genetic code to the amino-acid sequence in proteins; the former, despite being made up of only four bases, can code for 64 amino acids. However, only 20 amino acids are ever actually manufactured in living cells. Thus not all permutations of potential amino acids are exploited. The potential breadth of expression in the genetic code in other words is both channelled and filtered into only 20 amino acids.

By sequencing (analysing the sequence) of amino acids in humans and four other higher vertebrates (the horse, the cow, the pig and the rabbit) Zuckerkandl was able to calculate that the average number of differences of amino acids between their polypeptide chains was 22. It was known from classical palaeontology that the probable time that these four taxa had diverged from each other was 80 million years ago – in the late Cretaceous. Since there are two different types of polypeptide chain in this common form of the haemoglobin molecule, we can conclude that on the average there are therefore 11 amino-acid differences (or 'substitutions' as molecular palaeontologists prefer to call them) between the haemoglobin structure of the various polypeptide chains. Dividing this simple figure – which we can think of as average molecular distance – by elapsed time gives an estimate of the mutation rate that underlies observed changes in these respiratory molecules – in this case a rate of one mutation every 7 million years. With the mutation rate (or rate of evolution!) of haemoglobins thus calibrated Zuckerkandl realised that it was possible to ask a much deeper question: how much time would it take for all amino-acid substitutions to be accounted for? Or, to put it another way, how old, exactly, is the common ancestor of this most common molecule in the vertebrate family? The answer turned out to be about 380 million years old – placing the ancestral haemoglobin molecule fairly and squarely in the Devonian period – the time of the first fish – and therefore an entirely plausible estimate.

Zuckerkandl had proved that molecules could be used as measures of mutation distance and the speed of evolution. By the end of the sixties another, even more diverse, group of proteins was starting to be investigated. At the vanguard of this molecular offensive were Walter Fitch and Emanuel Margoliash, both working out of mid-western universities in the United States. In 1967, only five years after Zuckerkandl and Pauling had formulated the concept of semantides in the first place, Fitch and Margoliash published a now classic paper. Their approach focused on the cytochrome c's, a group of closely related proteins that are intimately concerned with the control of respiration in the cell and consequently are much more widely distributed in the living world than the vertebrate globins examined by Zuckerkandl. In theory, using cytochrome c's it is possible to extend the molecular phylogenetic technique all the way back to the time when oxygen use first became the common basis of respiration billions of years ago. Hand in hand with their exploitation of the cytochrome c's Fitch and Margoliash addressed the need to develop the mathematical techniques which were used to compare polypeptide sequences. In this they were aided by the arrival of the electronic computer which meant that their subtle mathematical programs could be executed with blinding speed.

This increase in speed meant that the investigation of relationships could be extended beyond animal and plant species to include the biggest and most enigmatic group of living organisms on the planet – the bacteria – and solve one of the biggest problems of molecular biology: their evolution and classification. This problem arises because bacteria have no distinguishing morphological features to speak of. Since the 1920s, various attempts had been made to classify them, but had always foundered upon this simple fact. These attempts revolved around classifying them according to their mode of nutrition or even the nature of the infections that a proportion of them induced! Such typological classifications have nothing to do with evolution, of course, they are strictly expedient, and by eschewing any attempt to understand homologous relationships they were doomed to failure. By the time of the Second World War, serious attempts to classify the bacteria had all but been abandoned; the problem was regarded as insoluble and swept under the carpet. On the face of it, the protein polymorphism technique presented a perfect solution to this problem.

But it was not long before the scientists working with this radical new technique began to experience difficulties which became steadily more acute as the 1970s progressed. The problem is that an enzyme, despite being a molecule constructed directly by the action of the DNA-RNA machinery, is still not the same beast as the genetic code itself. They are qualitatively different. They are tertiary semantides. The haemoglobins, for example, are functional useful molecules and as such are 'visible' to selective pressure – the driving force of evolution. Those that confer selective advantage are retained by evolution through the agency of differential reproductive success (Darwin's mechanism of natural selection): those that do not offer advantage are weeded out. In other words the amino-acid residue substitutions that accrue in the polypeptides that make up proteins do not truly reflect the random nature of mutations which occur in the genome. Such mutations may be random, but their expression in this higher tier of the seman-tophoretic pyramid is controlled by an external force; the protein mirror is not, after all, an accurate reflection of an organism's genome; the mirror is crack'd.

The reluctant palaeontologists

The 1970s saw the solution to this problem. A scientist who was interested in the enigmatic evolutionary relationships of the bacteria, Carl Woese, a young molecular biologist working at the University of Illinois, realised that it was pointless looking at proteins, as they could only ever give a distorted picture of true evolutionary relationships and tempos. By retaining selectively useful mutations, tertiary semantides effectively increased the true phylogenetic distance between species. On the other hand, he realised that truly 'free-running' semantides were of little value either. Experiments had showed that they changed too fast – they were an evolutionary stopwatch – which ran down quickly and thus was only useful for measuring rates and distances between closely related species. But Woese also realised that within the cell there existed a class of molecules that overcame both problems and was as intimately related to the sequence of nucleotide bases in the DNA as it was possible to get without actually measuring the DNA itself – the ribosomal RNA (rRNA) molecule.

To understand the importance of Woese's vision we need to understand a little of the function of RNA. RNA exists in three forms in the cell, all of which are involved in the construction of proteins from the information contained in the DNA molecule. The largest and most common form of RNA is rRNA which exists in the form of structures within the cytoplasm of the cell known as ribosomes. It works in conjunction with the two other types of RNA – messenger RNA (mRNA) and transfer RNA (tRNA). It is the ribosomes that are the *physical sites of protein manufacture* in the cell. The system works thus: messenger RNA (mRNA) is formed from DNA by 'transcription' (where the sequence of nucleotide bases on the DNA strand is transcribed into a strand of RNA). This mRNA then attaches to the ribosome where it becomes rRNA. Here the smaller, more mobile molecules of tRNA bring the ribosome amino acids which are then assembled into polypeptides and ultimately proteins. This latter process is 'translation'. Woese realised that rRNA held the key to deciphering evolutionary rates and relationships because in its large molecule it has segments that evolved both fast and slowly – the latter segments being the ones that tended to accumulate favourable changes under the influence of natural selection. We can use a metaphor to understand this: the rRNA molecule's usefulness lay in that it had both an hour and a minute hand. Woese also recognised that the rRNA molecule is incredibly ancient; it is intimately involved in the basic metabolic machinery of the cell and so should be able to illuminate evolutionary relationships all the way back to the very beginnings of life.

It was this antiquity that raised the possibility of an answer to the problem of the evolutionary relationships of the bacteria. It had been as early as the 1930s that a Frenchman named Eric Chatton had made the observation that a more natural division of the world was into those organisms that had their genetic material organised into a membrane-bounded organelle – the nucleus – and those that did not. In this latter (exclusively single-celled) group, the DNA floats free in the cytoplasm.

The former group, which Chatton called the Eukaryota (eukaryotes for short) includes single-celled organisms as well as more advanced forms where the cells clump together and exploit the advantage of cooperation (the metazoa). The latter group (the Prokaryota or more informally the prokaryotes) includes bacteria as well as certain bacteria-like organisms that are known as cyanobacteria (or blue-green algae because of the special pigments that they use to fix carbon from sunlight and carbon dioxide). The prokaryotes, too, loosely exploit the advantage of mutuality. In the present day they still form strange stone-like structures of calcium carbonate called stromatolites in tidal pools in Australia. These stroms (as they are colloquially known) dominated the early Earth of the Proterozoic aeon (the aeon before the Phanerozoic). Chatton's division was an early attempt to impose natural order on life on Earth by recognising that fundamental organisational differences at the cellular level were much more important than gross phenotypic differences.

Chatton considered the prokaryotes with their uncoralled genetic material to be more primitive than the eukaryotes and this view came to be supported by two scientists, Stanier and van Niel, who across three decades of collaboration investigated the chemistry and structure of bacteria to see if they really did represent a fundamentally different category of life. By 1962 they had concluded that Chatton's distinction between the prokaryotes and eukaryotes was not only accurate but if anything did not go far enough. They found that the distinction between the two groups was more profound than merely the absence of a membrane-bounded nucleus. Not only was there no membrane separating the genetic material from the cell, there were no membranes separating the organelles which controlled respiration and photosynthesis from the rest of the cell either. Furthermore, nuclear division occurred by simple splitting of the genetic material without the more complicated apparatus associated with eukaryotic cells. Finally, the cell walls of all prokaryotes was composed of a structural element – a mucopeptide – that was not known in the world of eukaryotes. So fundamental were these differences that it was impossible to avoid the conclusion that they went all the way back to the very beginnings of life on earth.

By 1970 Stanier and van Niel's distinction between the eukaryotes and the prokaryotes was the fundamental underpinning of molecular biology but had yet to make much impact on palaeontology. But throughout biology, spreading like ripples across a pond, was the clear recognition of the single huge implication of this duality, which was that the prokaryotes with their loose genetic material must therefore be primitive, and that the eukaryotes with their membrane-bounded material were correspondingly advanced. The importance of intercellular membranes had assumed iconic status amongst those who studied creatures of the living world.

Lynn Margulis was the chief proponent of a radical idea that revolved around two unique features found in different types of eukaryote: the mitochondrion and the chloroplast. Apart from their possession of membrane-bounded nuclear material, a universal feature of the eukaryotes is their possession of mitochondria – membrane-bounded organelles whose specific function is the control of respiration (energy production) in the cellular economy. Another feature of those eukaryotes that photosynthesise is their possession of chloroplasts, the organelle where carbon is converted to sugar from carbon dioxide using sunlight. Margulis' idea was that these two features of eukaryotic cells had not evolved *in situ* at all. Instead they were the result of an infection of one prokaryotic cell by another prokaryote sometime in the Proterozoic or Archaean. Margulis' theory was consistent with the tiny amount of data from the fossil record then available. This suggested that prokaryotes, indistinguishable from modern cyanobacteria ('blue-green algae') existed as long ago as 3.5 billion years before present while it was not until 1.5 billion years ago (at maximum) that the eukaryotes arose. The intervening 2 billion years was plenty of time for two 'infections' by mitochondria- and chloroplast-like prokaryotes to found the eukaryote lineage.

Margulis' theory became known as the 'endosymbiotic' theory of the origin of the eukaryotes and was published in 1970. The prokaryote/ eukaryote concept and the endosymbiont theory provided for the first time a firm theoretical underpinning for molecular biology. For the new science of fossils, the coincidence of time and concept was nothing if not fortuitous and it was not long before Woese brought his new technique to bear in testing these new underpinnings of biology. To begin with, all that Woese had to work with was the relatively primitive technique of splitting clumps of RNA into measurable units known as 'oligonucleotide sequencing'. rRNA molecules were split apart using an enzyme that homed in on the guanine nucleotide and broke the bond that attached it to its neighbour. The composition of the resulting mixture was then measured chromatographically. By comparing the oligonucleotide sequences from a wide variety of organisms – particularly bacteria – it was possible to build up charts, known as dendrograms, showing the relationships between the different organisms. But Woese appreciated that this approach was only a relatively minor advance on sequencing the amino-acid residues in proteins. You might be looking at a qualitatively different level from the proteins – the next step up the semantophoretic chain towards DNA itself – but you were still only looking at clumps of nucleotides, not the functional units themselves. The step forward was to sequence rRNA itself. This was done by using a then recently discovered enzyme called reverse transcriptase.

The exploitation of this technique was the breakthrough. Woese and his collaborators started extracting long chains of rRNA from the cells of as many different organisms as they could lay their hands on, including of course the group that had now come under suspicion as crucial to understanding the evolution of life: the bacteria. By comparing the number of differences between the sequence of RNA nucleotides in these different organisms, they were able to build up a map of the true genetic distances between organisms. They were also able, by calibrating these genetic distances against known dates of divergences in the fossil record, to start to compile data on 'true' (as Woese thought at the time) rates of evolution.

An early triumph of the Woeseian rRNA revolution was confirmation of Margulis' endosymbiotic theory. rRNA sequencing showed that both mitochondria and chloroplasts are descended from prokaryotes. It was a notable early success for the technique but it was not long before Carl Woese had compiled enough rRNA sequence data to stare two incredible truths in the face. The first of these suggested – no, shouted – that Chatton, Stanier and van Niel had not gone far enough. Their view of a fundamental cellular dichotomy in the tree of life was true enough – yet it was not two-fold, it was three-fold; the tree of life divided naturally into three branches.

Woese's preoccupation with the ancestry of the prokaryotes – in particular the bacteria – had proved the nemesis of the old order, for the rRNA nucleotide sequence data were unequivocal; the prokaryotes *themselves* divided into two fundamentally different groups on the basis of a suite of features that included the use of a completely different set of lipids (the basic constituent of fat) within the cell and a cell membrane based on a different set of long-chain molecules. But there was another crucial difference which separated these two groups of prokaryotes and persuaded Woese that both were worthy of fundamental distinction in their own right. The new group that he named the 'archaeobacteria', while retaining other fundamental prokaryotic features, had a system of enzymes controlling the transcription and translation processes that was more similar to that of the *eukaryotes* than to the remainder of the prokaryotes (which he called the eubacteria). Specifically he found that archaeobacterial RNA polymerase – the fundamental and crucial enzyme that controls the transcription of DNA into messenger RNA – was more similar to eukaryote RNA polymerase than to eubacterial RNA polymerase. Similarly, the protein structure of archaeobacterial ribosomes (where messenger RNA is translated into the amino acids that ultimately make up the cells' functional proteins) was more similar to that of the eukaryotes than to the eubacteria. It was this one essential difference that eventually persuaded most of Woese's colleagues that he was correct.

The really extraordinary thing about this revolution was that the biology of the archaeobacteria was almost totally unknown. Their unique and commanding position in the tree of life was worked out from their rRNA sequences alone. Prior to 1977, all that was known about them was that certain 'prokaryotes' favoured a rather outré lifestyle in extreme habitats. The archaeobacteria are a diverse and radical group that includes forms that require methane gas as the basis of their respiratory pathways and to whom oxygen is fatal; a group (known technically as extreme *halophiles*) that requires both oxygen and salt – the latter in very high concentrations; as well as the extreme *thermophiles* that live in hot springs where the temperatures can reach 110°C. Once Woese showed that they were a discrete taxonomic entity in their own right, their strange lifestyle was readily linked to their unique cellular architecture.

Following the Woeseian revolution, animals, plants and fungi (groups that as recently as the late 1960s were considered to comprise three-fifths of the diversity of life!) have now been relegated to insignificance in a remote corner of the eukarya.

But 1977 was a vintage year for Woese, for not content with introducing a new structure to the understanding of life on our planet he also dropped another bombshell, this time to do with the ancestry of the eukarya, our own branch of the universal tree of life. By the mid-seventies detailed analysis of rRNA sequences proved that the initial split between the prokaryotes and the eukaryotes simply could not be located. Woese eventually concluded that the eukaryotes had not split off from the prokaryotes but that the two separated from a common ancestor, which he called the 'progenote', some time in the Archaean, about 3.5 billion years ago. The lack of supporting rRNA evidence for prokaryotic ancestry of the eukarya was itself supported from an unlikely direction.

The eukaryotes have what can only be termed 'messy' genetic machinery, with large amounts of DNA that simply do not appear to have a function (this is colloquially known as 'junk' DNA) as well as particularly complicated mechanisms for transcribing DNA into RNA whose controlling genes are spread higgledy-piggledy around the genetic material. This is in marked contrast to the prokaryotes whose genetic machinery is so neatly arranged into functional packets that they have been given a specific name of their own: operons. Woese's discovery that the eukaryotes did not arise from the prokaryotes explained the reason for this fundamental difference.

In only two decades, Woese's rRNA techniques had provided three massively important findings for biology – and, as the new science of fossils was beginning to realise – palaeontology, too. rRNA sequencing had proved the validity of the endosymbiotic theory of mitochondrial and chloroplast origins, had demonstrated the threefold division of life, and had shown that the simplistic view that the eukaryotes were descended from the prokaryotes could not be correct. Carl Woese had single-handedly changed for ever the way that life was viewed. The power of the rRNA technique had been amply demonstrated by these hugely important discoveries. Like King and Hare before them the world seemed the oyster for the molecular palaeontologists. So keen were they – and so sure of their new approach – that they didn't hesitate to tackle the biggest question of all. The search for the universal ancestor – the great grandparent of all life on Earth – was on.

The universal ancestor

After 1977, everybody in the molecular palaeontology business knew what the ultimate goal of their new science must be; it could be nothing less than the philosopher's stone, the holy grail, identifying the common ancestor and the branching order of the three fundamental divisions of life. But there was no clue as to what form this ultimate genetic ancestor would take and the division between the archaeobacteria, eubacteria and eukaryotes was so deep that it was far from obvious what had diverged from what. Did the eubacteria come from the archaeobacteria and the eukaryotes, did the eukaryotes come from some common eubacterial and archaeobacterial lineage, or did the archaeobacteria come from a common eukaryote and eubacterial lineage? Because of the similarity in the organisation of transcription and translation architectures between the archaeobacteria and the eukaryotes, the prevailing suspicion was that the eubacteria would be found to have split off from the line leading to the eukaryotes and the archaeobacteria before they themselves had diverged.

As for the ultimate ancestor, the molecular palaeontologists eventually gave it a special name: the 'cenancestor'. The cenancestor's formal definition was sanctified by no less than the grand old man of molecular palaeontology himself, Walter Fitch (of cytochrome c fame) as, 'the most recent common ancestor to all the organisms that are alive'.

Woese himself had begun to wonder whether or not the cenancestor was not some free-living sub-cellular component from which all three domains had diverged separately. Such an ancestor would have to be something of fundamental importance that was common to all three of the groups. After much thought, he came to the conclusion that it was in fact nothing less than the ribosome, the cell organelle responsible for organising the transcription of the genetic code from DNA to protein via the RNA machinery. Such a simple sub-cellular structure would not itself experience evolution in the sense that we understand it. It would be too primitive to be subject to selective pressures because the two fundamental requirements for this type of evolution to work – a genotype and a phenotype – would not yet be present. Instead, this cell ancestor, this 'progenote', was an entity in which the difference between the genotype and the phenotype was still developing, probably, they speculated, by establishing the basic molecular functions by which life is recognised.

On top of this, identifying the divergence order of the three fundamental branches of the tree of life (let alone the probable genetic structure of the cenancestor itself – be it progenote or some more fully formed cell), was proving to be highly problematic using the molecular techniques. The essence of any molecular phylogenetic technique, be it based on metabolic molecules like proteins or more fundamental semantides like rRNA is the recognition of 'out-groups'. That is, if you want to know the evolutionary relationship of any gene (or gene fragment) from a particular organism then you need to have something to compare it with. The fundamental basis of the semantide technique is comparing one organism's protein or RNA sequence with another to tell how different they are. Clearly then, as the tree of life is descended in the attempt to compare bigger and bigger groups (phyla, kingdoms and ultimately domains) two things will happen: the size of the 'out-group' will grow larger while the number of potential out-groups falls. Even Woese's three enormous domains – which together encompass all life on earth – were calculable because there was always an out-group (a comparator) to balance the investigation. But what do you use as a sister group if you've identified the ancestral genome of the earliest living organism and want to know where *it* came from? By definition there is nothing to compare it with: the search for the universal ancestor cannot be conducted using the traditional methods of molecular phylogeny.

Woese did not solve this particular problem, but Naoyuki Iwabe working in the Department of Biology at Kyushu University in Japan did. To root the tree of life, he and his research group realised that if they could not sequence an organism's genome or part thereof, they could sequence a universal gene (if such a thing existed). To do this, they would have to rely on a strange quirk of life's genetic machinery that had been known about since the work of Zuckerkandl and Pauling in the early 1960s: gene duplication. They theorised that if a gene duplication (i.e. two copies of the same gene) could be found that was common to all life on Earth, then it must have arisen *before* the divergences that gave rise to the eubacteria, the archaeobacteria and the eukaryotes. It must have occurred in the cenancestor's genome, if not before. Iwabe and colleagues found two genes that fitted the criteria. A gene that controlled part of the respiration cycle of the cell and a gene that controlled part of the ribosomal translation process. Both genes are fundamental to the operation of life itself and both showed that the archaeobacteria and the eukaryotes are sister groups. It is the eubacteria that represents the earliest branch of the tree of life.

The Iwabe rooting published in 1989 was all that Woese needed to codify his view of life. Only a year later, in 1990, he elevated the eubacteria, the archaeobacteria and the eukaryotes to 'superkingdom' status. Specifically he suggested that the archaeobacteria and eubacteria – the two groups that together had made up the old prokaryota and which he had discovered using the rRNA technique back in 1977 – should join the eukarya at the highest level of the taxonomic tree and be known henceforth as the Archaea and the Bacteria. Together these three divisions have now completely eclipsed the old concept of kingdoms in the living world. Carl Woese named them 'Urkingdoms' or 'Domains', and it is the latter term that is in common usage today. Despite the fact that the process by which the three domains were arrived at spanned almost thirty years, 1990 was when the Woeseian revolution came of age, and we see it now as the fundamental watershed when the semantide comparison technique stopped being a sub-field of molecular biology and became a field in its own right, molecular palaeontology.

The chainsaw and the tree of life

In the decades since the Woeseian revolution, many other discoveries based on the molecular sequencing technique have supported his arguments, now total genome sequences of a number of archaeal and bacterial groups have become available. Such comparisons confirm the fundamental tenet of the Woeseian revolution that many of the genes concerned with transcription and translation are common to both the archaea and the eukaryotes and that these processes are performed similarly. Another discovery has been that although the archaea do not have cell nuclei – they are, after all, prokaryotes – under certain conditions their chromosomes assume the classic beads-on-a-string configuration known from the earliest days of cellular microscopy of eukaryote cells. Also, archaeal DNA seems to be associated with eukaryotic proteins called histones. The fundamental similarities between the archaea and eukaryotes are undeniable. However, one of the implications of the Woeseian revolution, with its early branching leading to the three domains of life, is that the only bacterial genes to be found in eukaryote cells would be those derived from the ancestral proteobacteria that supplied the mitochondria and those from the ancestral cyanobacteria that supplied the chloroplasts.

It's worth noting that even these genes are not found in all eukaryotes. Recent work has shown that the most primitive eukaryotes – those that diverged from the eukaryote mainline before the acquisition of these organelles (*Giardia* is an example) – do not have such genes. Another implication is that any transferred genes should be those that are involved in the control of respiration or photosynthesis (not the more fundamental cellular processes that would have been inherited from the ancestral archaean). This, however, is not what is found. Eukaryote nuclear genes often derive from bacteria and a good number of these serve functions other than respiration and photosynthesis, which are almost as integral to the basic architecture of the cell as transcription and translation. Another complicating factor is that bacterial genes have been found in archaea as well. An example of this is the strange and wonderful *Archaeoglobus fulgidus* which meets (in terms of its cell membrane lipids and the architecture of the transcription and translation machinery) all the criteria for membership of the archaea but uses a bacterial form of a particular enzyme for constructing these membrane lipids. It also has a number of bacterial genes that help it acquire energy and nutrients in one of the strangest habitats of all, undersea oil wells. The overwhelming implication of these strange cross-contaminations is that gene transfer does not only operate vertically but also horizontally between branches on the universal tree of life. In the last few years this phenomenon has been given the name lateral gene transfer (LGT). LGT means that single genes or even whole groups of them can be transferred across species barriers rather than being confined to passage from parent to offspring. The implications for the Darwinian picture of evolution are nothing less than staggering. LGT explains how the eukaryotes – supposedly evolved from the archaeal line – have in them so many bacterial genes. There is also reason to suppose that LGT has supplied more genes than have been retained; useful genes will be kept by the recipient cell while those that do not confer advantage (or which are positively harmful) are removed via the usual mechanisms of natural selection. Thus, mutation and genetic recombination are no longer the only suppliers of the raw material of evolution. Rather, variation can be supplied from co-existing organisms. The idea of LGT also potentially explains a long-standing mystery. Many eukaryote genes turn out to be unlike those of any existing bacterium or archaean; they seem to have come from nowhere. Particularly noticeable are the defining genes for two eukaryote features – the cell's own skeleton and its system of internal membranes. Mitchell Sogin of the Marine Biological Laboratory at the Woods Hole Oceanographic Institution and Russell Doolittle of the University of California at San Diego suggest that such genes have come from an unknown *fourth* domain of life via LGT.

It looks therefore as though mechanisms that are not unknown to us in the present day may also have been central to the evolution of the earliest cells on Earth. Think for example of the way – and the speed – with which bacteria can gain immunity to our arsenal of antibiotics. Is it possible that Woese is right once again and that the eukaryotes did not arise from the archaea at all? Perhaps they arose from some more enigmatic precursor cell – the progenote – that was itself the product of any number of earlier lateral gene transfers, events that made a cell part bacterial, part archael, and, if Sogin and Mitchell are correct, part something else. The progenote may have been radically different from anything that we know today. It is therefore no longer safe to assume that mitochondria and chloroplasts from alpha-proteobacteria and cyanobacteria were the only lateral gene transfers that occurred after the rise of the first eukaryotes. Only in later multicellular eukaryotes, after the advent of separated germ cells, can there have been restrictions on LGT. And so the evolution of sex represents not just a way of the cell increasing the amount of variation that it can produce but perhaps even a way of one form of variation outcompeting another. It is also entirely clear that bacteria and archaea have also indulged in the habit of extensive gene swapping. The new molecular palaeontologists, though, still cling strongly to the belief that rRNA as well as the proteins that oversee transcription and translation are unlikely to be easily moveable by LGT and that a tree of life that is based on them is therefore valid. But the non-transferability of even these genes is an untested assumption and such trees can be at best only a description of *part* of an organism's genome. A consensus tree, one that is based on more than one set of genes might be a better bet, yet such a tree would demand value judgements, for weighting would have to be applied to different gene sets and would therefore result in oversimplification. One that would not be much better than one based on morphology and certainly not what the original architects of molecular palaeontology – Zuckerkandl, Pauling and Woese had in mind.

So what does the true tree of life look like? At its top the crown of the tree probably continues to be correct in its depiction of the relationships between multicellular animals, plants and fungi and the gene transfers associated with the evolution of mitochondria and chloroplasts would still appear as fusions of the major branches. But lower down, below these transfer points, and continuing up into the modern bacterial and archaeal domains we would see a great many branch fusions. Deep in the realm of the prokaryotes and perhaps at the base of the eukaryotes the identification of any trunk as the main one would be entirely arbitrary. Even this revised picture might be too much of a simplification because branch fusion does not mean fusion of a whole genome. The full picture of the tree of life then would have to simultaneously display the superimposed genealogies of thousands upon thousands of different genes of which the rRNA genes are just one tiny part. Without LGT any one of the trees based on different groups – haemoglobins, cytochromes, rRNA – should all ultimately have converged on a single unique pattern: the history of life itself right down to the last common ancestor in which the ancestral genes of each tree would have been present. In fact, trees based on different molecules often do have regions of similarity. But extensive LGT means that we will never be able to reconstruct the genetic makeup of the common ancestor. Woese himself has put it thus, 'The ancestor cannot have been a particular organism – a single lineage – it was a communal, loosely knit, diverse conglomeration of primitive cells that evolved as a unit and eventually developed to a stage where it broke into several distinct communities which in turn became the three primary lines of descent which we know as bacteria, archaea and eukaryotes.' In other words, early cells (each having relatively few genes) differed in many ways but swapping genes freely meant that they shared various of their specific talents with their contemporaries and these cells eventually became diversified into the three major domains. These domains remain recognisable because these days LGT goes on *within* but not *between* them.

Some biologists are aghast at these notions because they strike at the heart of the mandate that Darwin gave us – to reconstruct the unique structure of the tree of life. But in fact the new science of fossils is working just as science should. The subdiscipline of molecular palaeontology is opening up new vistas and raising new questions that need answers. And it is using techniques that were undreamed-of in the days of the Victorian scientists who founded the new palaeontology.

The Organ Grinders

For all the success of the rRNA approach it is, like protein phylogenies before it, still only the monkey that plays around the feet of the organ grinder. It is not a direct look at the total genome itself. The advantage of the globins, the cytochromes and the rRNAs is that they are all relatively large molecules and highly conservative in evolutionary terms because of their fundamental metabolic importance. But as the film *Jurassic Park* showed clearly the aspect of molecular palaeontology that truly grabs hold of the public imagination is the idea of looking at the DNA of fossils themselves, and the possibility of using this ancient DNA perhaps to bring them back to life. This difference is a quantum leap from the rRNA work of Woese and his colleagues. What then is the truth of 'Jurassic Park' palaeontology?

The story starts in the only place such stories can start, in the hothouse atmosphere of the technology kings, California. Somehow it seems appropriate that the land of eternal sunshine – where America's boundless enthusiasm never dies – should be where the molecular palaeon-tologists decided that not only could they work out the tree of life but that they could, as well, indulge in more direct time travel. It was a classic example of the right idea in the right place at the right time. The time was 1984 and the place was the University of California at Berkeley, where

Allan Wilson, a molecular biologist, and his colleague, Russell Higuchi, published work that showed that DNA could indeed survive after death. They had studied the 'quagga' – a zebra-like member of the horse family that had lived in Africa before being made extinct by man in the middle of the nineteenth century. Samples from a 144-year-old specimen in a German museum yielded DNA from cellular mitochondria that allowed Higuchi and Wilson to show that the quagga was – as its exterior form already suggested – closely related to the zebra and only distantly related to the rest of the horse family.

The notion of reconstructing fossil DNA was surfacing elsewhere, too, in the unlikely setting of the small Swedish university town of Uppsala. In this quiet yet intense university atmosphere a young Swede was neglecting his thesis work in the early eighties, intrigued by the prospect of being able to reconstruct the DNA of dead organisms. Impressed with the ease with which DNA was being sequenced in living organisms, he found himself wondering how easy it would be to reconstruct the DNA of fossils. Rather than a recently extinct organism, Svante Pääbo focused on a lineage that was still very much alive, but some of whose members tended to go to great lengths to preserve their dead. Svante Pääbo contacted the curators of the University's Museum of Egyptian Antiquities and they provided him with small samples of skin and bone collected from 2000-year-old Egyptian mummies. Pääbo was obliged to indulge in some fast talking to acquire the samples since the museum's curators were less than enthusiastic once they learned that the extraction process was destructive. But they did cooperate, and even more importantly they had good contacts with a museum which was at that time behind the iron curtain in East Berlin. With a sterile scalpel Svante Pääbo removed a few grams of tissue from the twenty-three mummies that appeared to be best preserved, spending four days with the chief conservator of the state museums of Berlin looking through material some of which had been partly destroyed during the Second World War.

Back in Uppsala, Pääbo worked nights and weekends on the DNA problem while pursuing his thesis work during the day. Examining the specimens under the microscope, he found that they varied dramatically in their state of preservation. Most of the specimens were badly degraded, but there were a few exceptions. Peripheral parts of the bodies such as skin from the fingers and toes turned out to be better preserved. These tissues quite often contained cell nuclei that accepted chemical stains formulated to show up the presence of DNA. Pääbo reasoned that these cells retained their DNA because the processes that result in the destruction of DNA typically require water and samples from peripheral parts of the body would be more likely to be at least partially immune since they would dry out most rapidly. He worked as one would with any modern tissue, dissolving proteins with enzymes and using solvents to extract the DNA. Using electrophoresis (a modification of Abelson's chromatographic technique where an electric field is applied to help separate the molecules) he separated out the DNA fragments according to their size. The smallest fragments migrated furthest. The results showed that the DNA had been degraded to fragments only a hundred to two hundred base pairs long (in contrast fresh tissues contain DNA fragments that are more than 10,000 base pairs long). He first thought that such small fragments would be worthless but persuaded himself that even this small number was better than nothing: a wine glass may be half full or half empty depending on your point of view. So Pääbo proceeded to try to amplify the DNA that he had extracted. Amplification was the key, for fossil DNA is a rare beast at the best of times and to get enough to sequence, it must be copied.

The rarity of fossil DNA is in marked contrast to the situation faced by those who sequence DNA from living specimens where the pool is effectively limitless. DNA amplification in the early eighties was a tedious and heartbreakingly time-consuming task. It was achieved by bacterial cloning, by which a piece of retrieved DNA is inserted into the DNA of a bacterium where it is copied along with the bacterium's own DNA during its reproductive cycle. If this hybrid DNA (from which the term DNA hybridisation is derived) is introduced into a bacterial colony, then thousands of copies can be made. However, the process is still fundamentally one of addition.

Pääbo, finding that far fewer of the bacterial clones carried human DNA than he had expected, resorted to trying to isolate and replicate a common piece of DNA known as an ALU-repeat. ALU-repeats are sequences that occur nearly a million times in the human genome. He found an ALU-repeat in the tissues of a mummy that had been reliably carbon-dated as being between 2310 and 2550 years old. As Svante Pääbo was working on the mummy ALU-repeat, Higuchi and Wilson published the ground-breaking work which showed that DNA from the quagga was replicable. But it soon became clear why bacterial cloning was not producing the successes that molecular biologists had hoped for. The problem was that, whether you were dealing with a 2000-year-old mummy, a 150-year-old zebra relative, or a piece of week-old beef the average length of old DNA molecules was about the same because fragmentation occured in the first *hours* after death, before the tissues dried out. In addition, the presence of free oxygen radicals accelerates the process of DNA decay. In DNA hybridisation, the bacterial clones attempted to make copies of these damaged DNA strands thereby copying an erroneous template. Thus it was that in 1984 the progress of ancient DNA reconstruction hit a brick wall. But back in California, the answer lay literally just around a corner.

Revelation at Cloverdale. Junction of Route 128 and US 101, near Cloverdale, California, 38.46N, 123.01W. 15 April 1983

What is it about sports cars and the highways of California? Are they mystic ley-lines to some other dimension where a fast driver can slip back and forth through the gaps between chunks of reality? California's highways are where legends are born. Three hundred miles to the south of where the little car was climbing into the mountains of the Coast Range, James Dean met his maker just outside Lost Hills on Highway 466 and a generation was changed for ever. Further south still, Route 66 follows the track of the old Southern Pacific Railroad and retains a special place in the heart of motoring America as the final stretch on the old transcontinental route. And here, on Highway 101, although the driver of the little car didn't yet know it, the landscape of science was about to undergo a seismic upheaval.

But that Friday night in April 1983 Kary Mullis wasn't thinking about James Dean, or even the landscape of science. As the engine howled and he changed down for the curve just outside Lytton, and the wine-growing country of Sonoma and Napa fell away behind him, Kary Mullis was thinking of an experiment that he planned to perform for his employers on Monday morning.

Kary knew he was something of a maverick. As a graduate student in the biochemistry programme at Berkeley, he had written a piece of whimsy on the nature of time – with a grad student's facetiousness he'd called it 'The cosmological significance of time reversal' – and to his amazement it had been published by none other than *Nature*. Who ever said the Brits didn't have a sense of humour? He had received his PhD in biochemistry in 1972 and then had gone to the University of Kansas medical school to pursue post-doctoral work. But, although he picked up another wife in Kansas, he'd found that he missed California and so had moved back to the Bay Area for more post-doctoral work. But the writing on the wall was becoming more visible with each passing day; he'd been a post-doc twice and he could see that if he wasn't careful he'd be spending his days writing grant proposals to support himself, three kids and a wife. The prospect of a soft money existence – and on an academic's derisory salary to boot – had less than overwhelming appeal for a young man who was going places. It was enough to make a fellow seek gainful employment somewhere where his talents would be appreciated and appropriately remunerated.

At that time, the world of commercial biotechnology was unfolding like a bud in spring and small biotech companies were springing up all over California like chaparral shrub after a fire. Kary had seized his opportunity and had ended up at the Cetus Corporation of Emeryville, CA. He had spent the past four years making oligonucleotide probes. These are synthetic lengths of nucleotides that bind to complementary portions of a DNA molecule and if radioactively labelled can be used for determining whether a sample of DNA contains a specific nucleotide sequence or gene. But now in 1983 his interest in oligonucleotides was fast declining. The chemistry needed was no longer the black alchemy that had been such fun in the early days. Now it was being done reliably, quickly and boringly by machines. Automation had taken over here as in much else in science. Kary had found himself satisfyingly highly paid yet with time on his hands. He found himself wondering whether or not there was a faster technique for identifying genes than the oligonucleotide probe approach and it was this that brought him to the threshold of the Polymerase Chain Reaction (PCR).

Kary Mullis' big idea was based on a discovery made in 1955 by Arthur Kornberg of Stanford University and his associates on the nature of a cellular enzyme called 'DNA polymerase'. This enzyme serves several functions, including the repair and replication of DNA. It can lengthen a short nucleotide primer – a short section of DNA – which is complementary to a longer section by attaching an additional nucleotide to the free end. Since this nucleotide has to be complementary to the base in the corresponding position on the template strand it is possible to work out what that base must be. For example the A base will only pair with the T base and a G base will only attach to a C. By repeating this process, it is in theory possible to extend an oligonucleotide primer all the way to the end of the complementary DNA strand. A further advance was 'dideoxy sequencing', or the 'Sanger technique'. The Cambridge don who invented it has now been immortalised; the human genome project is based in the building that bears his name. It was Sanger who invented DNA sequencing. Sanger's technique makes use of a special type of DNA base known as dideoxynucleotide triphosphates (ddNTPs for short). Like ordinary nucleotides under the influence of DNA polymerase, these will attach to an oligonucleotide primer provided that there is a complementary template strand. Unlike ordinary nucleotides, however, the ddNTPs will automatically 'cap' the end of the primer and prevent the addition of any more bases. The Sanger technique allows molecular biologists to extend primers selectively with ordinary nucleotides and then stop the reaction at any desired length by adding the appropriate ddNTP. By arranging the resulting DNA fragments according to their length and knowing what ddNTPs have been added, an investigator can determine the sequence of bases in a template strand. This is the essence of what we now call DNA sequencing.

All this ran through Kary's head as he approached Cloverdale on Highway 101. The air was sweet with the heavy scent of the California buckeyes lining the road as it wound further and further up into the high lands, and a gentle, moist breeze blew inland from the ocean. As Kary approached the turning, his thoughts turned to the problem of potential contamination in Monday's experiment. From his work at Wolfgang Sadee's laboratory at the University of California in San Francisco, just before he had left the academic life to join Cetus, he remembered that there was a good probability that there would be stray nucleotides in the mixture. These nucleotides would complicate the interpretation of the electrophoresis gel and Mullis was wondering how he would get rid of them before performing the experiment. He considered and discarded the idea of using an enzyme – alkaline phosphatase – but realised that he could use DNA polymerase itself to remove the contaminants. By running the sample through a preliminary reaction which included the oligonucleotide primers and the DNA polymerase, but before adding the ddNTPs, he reasoned that he could deplete any nucleotides in the mixture by incorporating them into the extending oligonucleotides and then, simply by raising the temperature of the sample he would be able to separate these extended 'mock' oligonucleotides from the DNA targets. They would still be in the sample but because there would be more unextended primers than extended ones the DNA targets – when he actually came to perform the experiment – would be most likely to hybridise with the unextended primers as the mixture cooled. He could then add the ddNTPs and more of the DNA polymerase to perform the sequencing experiment proper. But although the aesthetics and economy of the experiment appealed to him, he still worried about the presence of the extended oligonucleotide 'contaminants' in the mixture: would these oligonucleotides interfere with the subsequent reactions when he tried to identify the unknown base? If they had been extended by many bases would they not perhaps be extended enough to create a sequence that included a binding site for the other, unextended nucleotide (the primer)? Then: this would not cause a problem! Far from it! The strands of DNA containing the target and the extended oligonucleotides would, of course, have the same base sequences. The mock reaction – envisaged initially as ridding the mixture of contaminants – would have the effect of *doubling* the number of DNA targets in the sample. Mullis' familiarity with computers and particularly iterative algorithms, portions of software that perform the same calculation on the products of the previous cycle, immediately made him realise the importance of the idea. This was a process that if reiterated would result in *geometric* increase in the amount of target DNA in a sample. Stopping the car, he pulled out a pad and pencil from the glove box, ignoring the complaints of his girlfriend. A few seconds scribbling and he had also realised that the length of the new strands would be naturally self-limiting because of the presence of so-called 5' primers. Consequently, he could replicate larger fragments simply by making primers that hybridised further apart on the source molecule. The fragments would always be discrete entities of a specified length. It was hard for Kary not to believe that such a beautiful idea had not already been discovered. But a literature search the following Monday – after a restless weekend during which he did nothing but consume quantities of Anderson Valley cabernet and worry – turned up nothing. And yet, over the next few weeks the idea seemed curiously unable to ignite enthusiasm in his colleagues. It took several months for him to prepare for the first test of the reaction. He used all his biochemical skills to make educated guesses about the buffer solutions to use, and in what concentrations and quantities, as well as how much to heat and cool the mixtures, as these molecules were exquisitely sensitive to temperature. He checked and rechecked Kornberg's early papers. When everything was ready he placed a virus and a couple of short oligonucleotide primers in a single test tube with the correct reagents. This was the moment of truth: would the short sequence of DNA bases that he had bracketed with the primers truly increase in number? The answer was an unequivocal yes. From there it was only a matter of talking to the Cetus patent attorney. Over the following months Kary confirmed that PCR would work on even hugely complicated human DNA and that it could be used to amplify a single gene.

Today, the polymerase chain reaction is the stuff of legend. There have been technical improvements; for example, the original DNA polymerase that Mullis used has been replaced by DNA polymerases taken from thermophilic bacteria, members of Carl Woese's beloved archaea. These enzymes are heat-tolerant and therefore do not have to be replaced in the solution after the heating step when the newly formed DNA strands are separated to add more primer. These days, high-temperature DNA polymerase is produced commercially by genetically engineered bacteria. No one was prepared for a process that could provide limitless amounts of DNA. After the years when bacterial cloning of DNA was a time-consuming process that resulted in at best a few hundred or a few thousand copies, the idea that rare DNA could now be magnified until it was effectively available by the bucketload took some getting used to.

And so by one of those coincidences that make science possible, at the same time that Svante Pääbo and Allan Wilson were stymied by the deficiencies of bacterial cloning for replicating ancient DNA, almost on Wilson's doorstep Kary Mullis was showing the way forward. DNA – as we have seen – is often damaged, and the bacterial cloning technique merely replicates the DNA as it is found, warts and all. But because Kary Mullis' technique would only work on intact DNA it intrinsically allowed one to isolate and 'amplify' (as the process came to be called) only 'clean' sequences.

Pääbo immediately realised the implications of PCR for the science of palaeogenetics. As he himself has said, prior to the discovery of PCR the study of ancient DNA barely qualified as a science. Because each single strand when accompanied by its correct primer would produce a classic double helix, the desired DNA sequence would double in quantity with each cycle. The selectivity of Mullis' chain reaction was another plus. If only one molecule survived intact in an ancient tissue, it would still be amplified by PCR. Damaged molecules could not dominate the sample because they were corrupted and therefore no longer amplifiable; they would not disturb the experiment. It was these potentials inherent in Mullis' discovery that brought Pääbo out to California to join Allan Wilson's team. For Pääbo, the allure of ancient DNA by now far outweighed his interest in his thesis work. What heady days those were! To be part of the birth of a new science and in California of all places; it was only right that Pääbo should journey to this biotechnology Mecca and assist in its parturition.

There was another Californian who saw the potentials as well; fearsomely bright, always on the look out for the next big idea, he had made his name with a string of hit science and technology thrillers. Michael

Crichton had started his career as a doctor but had given that up in 1969 after two incidents that had brought his life into sharp focus. The first of these was selling the movie rights to his hugely influential first novel *The Andromeda Strain* and the second was a brush with multiple sclerosis. He had realised then that life was too precious to spend on anything other than what one really wanted to do. Crichton was a man who, trained as a scientist, could take and understand complex scientific concepts and turn them into best-selling fiction. His speciality was and is being in the right place at the right time and recognising a good idea when he sees one. Crichton saw the potential of PCR and the rest is cinematic history. In *Jurassic Park*, Crichton took the potential of PCR and extrapolated from it in a controlled way. As always, the science in Crichton's writing is compelling and believable even to other scientists. For example, he envisaged a situation where a source of uncontaminated dinosaur DNA was available: blood from a fossilised Cretaceous insect! From that the Polymerase Chain Reaction amplified the sections of intact DNA, again perfectly believable. And from there he merely grafts on to the story – via a couple of Cray computers – the ability to identify the missing sequences of DNA and the use of another source of nucleotides – amphibian DNA – to fill in the gaps. Crichton's book was based on a revolution that was happening on his own doorstep.

After Pääbo arrived in Berkeley they used the PCR technique to reanalyse the quagga's DNA. There had been some concerns about the original bacterial cloning approach, particularly the fact that two sequences of nucleotides known to be common to all vertebrates had not been found. The PCR experiment showed that both these sequences were present and in the correct positions, as expected. The bacterial cloning technique had simply replicated an error that had been present in the original strand. This was to become a general finding with respect to PCR versus bacterial cloning. Because bacterial cloning starts with a single sequence, any error introduced by post-mortem damage is replicated ad nauseum. In contrast, PCR effectively starts with at least a few dozen nucleotide sequences, amplifying these to vast quantities and so any errors in single original strands tend to be cancelled out by whole and equivalent sequences in other originals. Another consequence of the PCR technique that Pääbo and Wilson discovered early on was that not only would PCR not amplify damaged DNA, it could actually *reconstruct* intact DNA from several partially degraded originals. The two PCR primers bind to the template and then extend either to the damaged site or to the end of the template molecule. In the heating step of the next cycle these molecules separate leaving extended primer molecules that are not damaged where the previous templates were (to use the photocopying analogy of the previous chapter, they are more nearly a complete page so that the next copies will be better still). So the process of piecing such strands together is effectively an automatic consequence of PCR. It follows therefore that under favourable conditions PCR can amplify sequences that are longer than any of the original templates. But, as work progressed, a problem with the PCR technique became apparent: the ease with which contaminating DNA could be copied. Contaminating DNA can potentially come from many different sources: museum curators or scientists as they handle the samples, or even the minute amounts of DNA from earlier experiments conducted in the same room. DNA is all around us, unnoticed until amplified a millionfold by the power of the polymerase chain reaction.

By the end of the eighties, another major advance in ancient DNA technology had been realised. Previously, it had always been thought that DNA would only be extractable from soft tissues, and since such material is vanishingly rare in the fossil record, ancient DNA studies would have to be pretty much confined to 'sub-fossil' relics, effectively material from archaeological rather than palaeontological timescales. But in 1989 Erica Hagelberg and Brian Sykes of the University of Oxford announced that they had successfully separated DNA from bone. For a time it was even thought that skeletal remains might be a better source of DNA than soft tissues, because DNA binds so easily to the calcium phosphate in bone and because bone is less susceptible to immediate post-mortem water damage. The discovery that DNA was extractable from bone also immediately raised the question of how far back one could extend the ancient DNA record.

But at the same time as this potential triumph loomed, bad news faced Svante Pääbo. Both the original quagga and mummy papers (published in *Science* and *Nature* respectively) had been accompanied by fulsome commentaries about the limitless vistas for the new science of fossils that they opened up. *Nature* had even put a photo of Pääbo's mummy on the cover. But for Svante Pääbo this victory was suddenly pyrrhic. He had gone to California to verify the sequence of the quagga using the newly invented PCR technology and they had been successful, showing that the two suspicious missing sequences were in fact where they were expected to be. But in 1989 two scientists, Giovanna del Pozzo and John Guardiola, showed that the DNA fragments from the mummified one-year-old boy that Pääbo had cloned was almost certainly a contemporary human contaminant. It was a personal blow for Pääbo but this news could not dampen the scientific chain reaction that was now underway in the new palaeontology. Excitement at the potential of salvaging fossil DNA was reaching near hysterical levels with more and more labs setting up to exploit the new techniques and only a year later the million-year barrier seemed broken. DNA was apparently extracted successfully from a Miocene magnolia leaf preserved in the clay-like sediments of a freshwater lake in Idaho. So well-preserved was the leaf that detailed tissue structures – even the cell walls and the organelles – were visible suggesting that the chances of amplifying intact DNA were high. This was a ground-breaking study. Previously, the most ancient material that had been examined were museum-preserved specimens of the recently extinct Tasmanian wolf *Thylocinus cynocephalus* and human brain tissue of archaeological age recovered from peat bogs.

Following the magnolia study and in less than a year, another group had sequenced apparently intact DNA from more plant tissue from the same locality in Idaho. But by now Svante Pääbo and Allan Wilson, the architects of the original fossil DNA studies, themselves had grave doubts about the validity of the technique. They found that they could not replicate the magnolia DNA sequence, raising misgivings about its authenticity. Pääbo, perhaps embarrassed by the problems associated with his original study, soon became a sceptical zealot. He had come to realise that the power of the polymerase chain reaction was at the same time its downfall.

It could amplify the tiniest fragment of DNA; a few nucleotides shed from a scientist's skin cell would be all that it took, and after a few cycles of the Mullis miracle you would have a measurable quantity of DNA! Pääbo saw that, unless everyone was very careful indeed, the credibility of the new science of ancient DNA would be destroyed. He made strident demands for the imposition of stringent laboratory controls to minimise the chances of contamination and was largely ignored, the excitement engendered by breaking the million-year barrier was too great. The genie was out of the bottle and would not be put back. 1990 was the beginning of the golden age of ancient DNA palaeontology, for it was the dawning of the era of amber. It was an eerie example of science-fact following science-fiction. Between the publication of Crichton's book in 1990 and the opening night of the movie in 1993 no fewer than three papers were published documenting the successful recovery of ancient DNA from insect material preserved in amber! In 1992 alone two separate groups claimed to have successfully extracted DNA from amber-preserved insects (a bee and a termite, respectively) of Oligocene and Miocene age.

The pace subsequently quickened with DNA sequences claimed from amber-entombed spores and even bacterial sequences. Once again, and like the periodic impact hypotheses of a decade before, public interest was intense and it was that (and the media's hunger for headlines) that was driving the science. But there were also sound reasons, as Crichton had correctly reasoned, for believing that extracting DNA from amber was possible. If geologically ancient DNA was to be found anywhere, amber would be the most likely place because the rapid entombment process minimised the time available for water- and oxygen-induced damage (hydrolysis and oxidation). The tissue would effectively be mummified. In addition, those components of the amber which impart its distinctive resinous smell would in theory inhibit microbial decay. On top of this, the preservation of amber-entombed fossils is normally exquisite, with structures all the way down to sub-cellular organelles commonly preserved. Biochemical preservation was also apparently excellent with amino acids retrievable in quantity and quality as good as that from specimens of archaeological age. At first sight, the DNA sequences from the amber-entombed insects seemed authentic; they even seemed to make evolutionary sense. The bandwagon was rolling so fast by now that it looked set for lift-off. Take-off velocity was achieved in 1994 when Scott Woodward, a geneticist working at the well-funded Brigham Young University in Utah, claimed that he and his co-workers had sequenced DNA from two large bones found in a local coal-mine. The bones had been found embedded in upper Cretaceous coal; the remains of plant fossils from a coastal swamp. The coal was 80 million years old. The bones were those of dinosaurs! The science of *Jurassic Park* had come full circle.

This was not a coincidence. As a boy, Scott Woodward had often sat in the fossilised dinosaur footprint his granddaddy had found. It was this thought – amid the media feeding-frenzy surrounding the release of *Jurassic Park* – that had given him the idea of approaching Mark Bunnell, a mining geologist friend in their home town of Price, Utah and asking him to call if they found any unusual bones. The call duly came and Woodward and co-worker Nathan Wey set about preparing the bone for DNA extraction, amplification and subsequent sequencing. The bone did not appear to be typically fossilised. It was yellow and friable and structures could be seen that were reminiscent of the osteoblasts, bone cells. There was no evidence of recrystallisation, as would have been expected had the bone been exposed to mineral-saturated ground waters. Structures in the bone stained positive with dyes used to identify the presence of nuclei in modern tissues. Yet the polymerase chain reaction was attempted almost three thousand times on extracts from the two bone-fragments before yielding just nine short fragments of a gene apparently from a mito-chondrion. These fragments proved difficult to match with any known genes but the sequences were plausibly intermediate between those found in reptiles and mammals. Hence the Woodward team speculated that the sequences were indeed degraded dino-DNA. But the rest of the scientific community were suspicious. As at the magnolia site in northern Idaho the environment of deposition did not seem likely to have been one that favoured the preservation of DNA. All the water that had been around at the time of deposition would certainly have been likely to promote hydrolytic reactions that would have degraded the DNA even if conditions thereafter were oxygen deficient. Also, the grade of the coal suggested that the bones must have been buried to a depth of several kilometres and subjected to temperatures as high as 95°C. The poor preservation of the amino acids in the bones also suggested that DNA would have been unlikely to have survived intact.

More recent reanalysis of the Woodward *et al* material has shown the DNA to be mammalian and almost certainly derived from a human 'pseudogene' sequence (mitochondrial DNA that has been incorporated into the DNA in the nucleus). So although Woodward *et al* had taken stringent precautions it appeared that they had reported a human contaminant after all. Had the excitement of the new technique seduced them and rushed them into print too quickly? With hindsight additional fossil bones should have been found and analysed. There was one further paper claiming successful sequencing of ancient DNA – from bacteria in amber – but the dino-DNA debacle effectively marked the end of the spectacular claims for successful retrieval of ancient DNA. After 1995, the 'ancient DNA' community imploded into a mass of squabbling and humiliated researchers who scrutinised and re-scrutinised each other's much-hyped results and found most of them wanting. One researcher even commented that in his opinion 85 per cent of the studies published in that 'golden' period were later shown to be flawed.

It is hard to blame the ancient DNA palaeontologists for having had their day in the sun; the temptation to bask in Crichton's reflected glory must have been too great to ignore. The problem boiled down to inadequate authentication procedures during extraction and amplification. Researchers were looking for sequences that made sense in terms of the evolutionary history of the animal or plant that they thought they were analysing. This in turn meant that they effectively had preconceived notions of what they were looking for. Any *other* DNA that was found (for example human DNA) was dismissed as an artefact and indeed was often not reported. Such was the power of the polymerase chain reaction though, especially in dealing with damaged templates, that sooner or later, via a series of erroneous nucleotide substitutions, it would construct a sequence that looked superficially like the one the researcher was searching for;

insect-DNA, dinosaur-DNA, or magnolia-DNA. Give monkeys typewriters and sufficient time and eventually they'll come up with *Hamlet*. Only with PCR, the time required was hours or days, not millennia. Reapply the technique to amplify the new strand and you would have a self-fulfilling prophecy. All this and more was hashed and rehashed at countless conferences in the mid-nineties. And then in 1997 as the dust settled one group working in Svante Pääbo's new ultra-clean facility in Munich, employing the most stringent anti-contamination protocols ever, successfully sequenced DNA from Neanderthal man. Methodologically, the Pääbo group used a different approach, determined that they would not be caught out again. The template molecule was first cloned, then sequenced and then these sequences were examined for portions that *differed* from those in living humans. These sequences were then used to design Neanderthal-specific primers that *could not* amplify modern human DNA. Using these specific primers, the resulting Neanderthal mitochondrial DNA sequences were then amplified and shown to be distinct from that of modern humans. Success at last!

One important difference was that a separate laboratory was used to double-check the results. The study concluded that the Neanderthals were a separate hominid lineage which did not share genes with the lineage that led to modern humans. Neanderthals, named from the German valley where the fossils were first discovered, are known from palaeontological evidence to have been about 30 per cent larger than modern humans and of enormous muscular strength. They had low foreheads with protruding brows and large noses and apparently ate meat. They became extinct about 30,000 years ago having lived in Europe and western Asia for at least 100,000 years. Towards the end of that period, Neanderthals co-existed with our direct ancestors (who may have been the cause of their extinction). The question that interested anthropologists and which this DNA study addressed was whether the Neanderthals were a race of ancient human beings which fell within the genetic range of modern humans and contributed to the modern gene pool or whether they were a distinct species: *Homo neanderthalis*. The DNA study retrieved a sequence of 378 base pairs from a specimen recovered 140 years ago in the Neanderthal valley itself. Within the DNA length investigated they found twenty-seven differences from the human sequence and these differences occurred at sites that tended to vary between modern humans and chimps. The data showed that the Neanderthals were closer to humans than to chimps but that the Neanderthals were still far from the natural range of human genetic variation. Indeed the races of modern-day *Homo sapiens* vary from each other by only eight nucleotide differences. Chimps and man diverged about 4 million years ago while the last common ancestor of *Homo sapiens* and *Homo neanderthalis* lived only about 600,000 years ago (probably in Africa). When the new Neanderthal data were incorporated into the evolutionary tree of modern *Homo sapiens* the result strongly suggested that our species originated in Africa too but only 150,000 years ago.

The key that allowed the success of this study after so many failures was the construction of Neanderthal-specific primers that would not amplify the DNA of modern humans. The biggest risk of contamination was therefore factored out of the equation early on. Complimentary amino-acid studies of the Neanderthal bone showed that the level of water-induced decay of the molecules in the sample was low; suggesting that the DNA was unlikely to be corrupted either. The Neanderthal DNA study may be the most important recent study, but several groups have reported the successful retrieval of mitochondrial DNA from 50,000- to 100,000-year-old mammoths discovered in the Siberian permafrost. This environment with its low temperatures and extreme dryness is excellent for preserving ancient DNA. These DNA sequences turned out to be very similar to those of Asian elephants, an observation supported by the mammoth's excellent fossil record.

But what of the fiasco of amber preservation of ancient DNA? In the midst of all the recriminations over reproducibility, a careful investigation was conducted by a group based at the Natural History Museum in London. The museum has a very large collection of amber-preserved insects, so a higher number of specimens could be examined than in all the previous studies combined. The group used several different extraction methods and PCR conditions as well as a variety of amplification protocols to test whether DNA really could be preserved in ancient amber. They discovered that occasionally a sequence could be amplified apparently independently of the presence of a fossil insect! Where an insect *was* present any DNA sequences successfully amplified were unrelated to it. Perhaps the most important part of the work that the group carried out – and certainly one that should have been done before all the other studies had rushed into print – was the investigation of much younger insects preserved in copal (what pine resin initially solidifies into before it transmutes into true amber). They found that such insects were useless as potential sources of amplifiable DNA, and that therefore every previous report of the recovery of ancient DNA from insects in amber could all be disregarded as experimental artefacts. Cooler contemplation had also reminded many workers that amber is gas-permeable so that any DNA, although removed from the risk of water-induced degradation, would not after all have been removed from the effects of oxygen-induced breakdown.

What then is the future of ancient DNA research? The number of papers on ancient DNA has declined over the past few years as has the number of reports appearing in high-profile journals. In the first decade of ancient DNA research, 35 per cent of articles appeared in *Nature, Science* or the USA's prestigious *Proceedings of the National Academy of Sciences*. In the last three years this proportion has declined to 15 per cent, a figure comparable to that of other papers in the palaeontological or biological sciences.

The honeymoon period of ancient DNA research is now well and truly over but this is no bad thing. More careful work is needed and exciting results are anticipated. However, as far as *Jurassic Park* science is concerned the harsh truth is that it is unlikely that any DNA is preserved in material more than 100,000 years old.

And the dinosaurs died out 65 *million* years ago.

CHAPTER TEN

Battles of Deepest Time and Space

In Jules Verne's epic, Journey to the Centre of the Earth, the travellers return to the surface of Earth via a volcanic vent that eventually expels them on to a Sicilian hillside with the blue mountains of Calabria towards the east and the fiery Mount Etna erupting nearby. On their way back to the surface, they rise through the ages of life – the Phanerozoic aeon. Imagine now that you are at the foot of that volcanic shaft but instead of there merely being one way out – up, the route Verne's bold travellers took – there is another, dark-mouthed tunnel leading off to one side and inexorably downwards. What would you do? Vote to go up and home, or continue your journey to the centre of Earth?

The future of the new science of fossils lies through that dark opening, down and away to far vistas of uncharted knowledge, down to the bacterial beginnings of life and away, out beyond Earth to other planets and solar systems. The new science of fossils has only served its apprenticeship in the Phanerozoic. Its real challenges now are both older and further away.

The mouth of that dark opening represents both a specific time in the history of life and simultaneously a confluence where all the pathways of modern palaeontology meet: a time horizon where the techniques and efforts of all those architects of eternity come together to create something that is greater than the sum of its palaeontological parts. To understand this Cambrian explosion and the events that preceded it is to address one – but only one – of the biggest questions of modern-day palaeontology. The protagonists in this battle of deepest time deploy many of the techniques that together form the new science of fossils: cladistic analyses to objectively map relationships between organisms; dates which underpin necessary timescales based on the subtle fluxes of the radiogenic isotopes; the state of the ocean and atmosphere using stable carbon isotopes. The reluctant palaeontologists, those molecular alchemists, have their role to play in this drama too and, as I write, are becoming more vociferous in the debate about the so-called 'slow-burning fuse': just when, exactly, were the elements required for the explosion of multicellular life at the base of the Cambrian formed, and why is the fossil record of these events so poor? With a certain pleasing symmetry, the debate about the Cambrian explosion has its origins in the same era as Marsh and Cope and Elles and Wood, broadly at the turn of the nineteenth century. Since then it has rumbled on, gathering speed and invective all the way through the twentieth century until now it stands as one of the biggest debates of the new science of fossils . . .

If I could talk to the animals . . . Burgess Pass, British Columbia, 51.20N, 116.51W. 30 August 1909

This high the air was thin to the point where it would be hard for most people to breathe. But Charles Walcott barely noticed; this was not, after all, the first time that he had collected in the high Rockies. His interest in the formations of north-west Montana dated back to when he first crossed the Belt Mountains in 1900. He had been collecting on and off in Montana and Idaho ever since but these rocks of the lower Cambrian extended into Canada, tempting him sorely to cross the border and extend his work. In 1907, he had been appointed Secretary of the Smithsonian Institution in Washington DC, a position with more prestige and power than he had ever had as director of the US Geological Survey, and he had used the change as an excuse to move his field area further north. The pleasure of riding the Canadian Pacific with his family, leaving behind the crushing burden of administration, was still undiminished even after two years. This was his bolt-hole, the place where he, Helena and the children could come and geologise, where the bureaucratic monster of Washington scientific life could not easily find him. But the field season was nearing its end and no matter how he tried to avoid it the thought of his imminent return to the capital would not go away. In a month, or probably less, the snow would come and within hours all the localities that he had so painstakingly sampled over the summer would be covered over. A white blanket that prevented any possibility of access or sampling and yet would also serve to keep their secrets safe from competitive eyes. In less than a week the family would need to return down the mountain to the station at Field and head back East. He shook his head, suddenly determined not to let the thought spoil these last days; for the moment he would simply enjoy the view. Behind him towered the sharp peak of Mount Wapta, to the east was the steep scree-covered slope of the mountain, to the west was Emerald Lake and the ranks of the President Range stretching to the south and the American border.

Suddenly, as he rounded a corner on the trail, he noticed a loose block. Dismounting, he examined it. On its surface – he could scarcely believe it – was a silver tracery outlining several phyllopod crustaceans. Several minutes later he sat back on his heels in disbelief. Every surface of the block was covered with some of the finest arthropod fossils that he had ever seen. Standing and shading his eyes he looked up. A hundred feet above him the seventy-degree slope became a vertical face, which he recognised as a part of the Stephen formation that they had been collecting for the last several days. They had found several interesting fossils on the west slope of the ridge that stretched between Mount Field and Mount Wapta but none had yielded specimens as good as this. The aspect of the fossils intrigued him. The silvery sheen was unlike anything that he had ever seen before and there seemed to be a depth to the specimens as though they were more than compressions on a bedding plane. The obvious place for them to have come from was the vertical cliff a hundred feet above him and he noted its position in the last rays of the dying sun before he went back down to join the others at camp for the night.

Early the next morning, they went back together and Walcott sketched in his notebook pictures of the fossils as well as the position of the vertical face from which they must have come. Over the next few days he drew assiduously and took several panoramic shots with the heavy camera that young Rutter lugged up from the camp. Within a couple of days he and Rutter had narrowed down the bed from which the fossils must have originated to one that contained a fine group of sponges apparently *in situ*. For the next several days he and the family worked on the outcrop and every day brought new and more wonderfully preserved material. This was more than he could ever have imagined. In his years directing the Geological Survey, establishing the Geophysical Lab (even against Carnegie's wishes!), and then taking on the secretariat of the Smithsonian nothing had been able to surpass the simple joy of discovery. And nothing in his long palaeontological life had ever been of this magnitude. It was almost more than he could bear when the week was up. On 7 September, with his son Stuart and young research assistant Rutter scrambling close behind, he had to acknowledge that this must be his last day on the mountain. At least for the year 1909. In his mind, though, a field trip for 1910 was already a certainty. He barely admitted it to himself but the pleasure of fieldwork was such that, even without this find, he'd have surely come anyway.

Charles Doolittle Walcott qualifies – if anybody does – as an architect of eternity. He was born on 13 March 1850, the son of the owner of a knitting mill in a small town just west of Utica in upper New York state, an area that would subsequently become well-known for its fine Palaeozoic invertebrates. It was a classic case of the area making the man. When he was only two years old, Charles Walcott began picking up fossils from the Utica shale, a rock formation now known to be part of the Ordovician system, a system that had not even been proposed by Lapworth at the time Walcott was born. When he was thirteen, he spent a summer at Trenton Falls, New York collecting from the Trenton limestone. His interest in fossils had grown into a passion. His uncle, who had brought him up, wished him to study for the ministry, but Walcott had no such interest and declined. Instead he went to work in a hardware store until he was twenty-one. Having had enough of clerking, he decided that he would try to make his passion for fossils pay. Returning to Trenton Falls, he spent the next five years with William Rust, a farmer who shared his interest in palaeontology. In return for his keep, Walcott helped out around the farm but also did a lot of collecting. The fossils of Trenton were a profitable sideline for both Rust and his young assistant. In 1873, Walcott and Rust sold their collection to the Harvard savant Louis Agassiz for $5000, an incredible sum in that day and age (the practice of selling fossils for profit is not a new phenomenon!). The relationship with Agassiz was more than a commercial one, for an inspired Walcott decided that there was more to the study of fossils than cash. He wrote his first paper in 1875, a description of a trilobite from the same formation in New York state. This, though, was a bad time for Walcott; his wife, one of Rust's sisters, died after only slightly more than a year of marriage. Walcott was devastated. Two years later, James Hall of Albany offered Walcott a position as a research assistant. Walcott received $75 a month in this, his first paid position as a professional palaeontologist. He was the only one of Hall's assistants who published under his own name and while working for Hall, Walcott was the first palaeontologist to describe the limbs of trilobites. But Hall was jealous of his brilliant young companion and it was not long before the two fell out. In 1878 Hall decided not to renew Walcott's contract. Despite his lack of employment, Walcott continued to study trilobites and to write about them in Albany. But after a year he returned to Trenton Falls. That same year the United States Geological Survey was founded with the great explorer of the West, John Wesley Powell, as its first director and quite unexpectedly in July, only three months after the Survey's articles of incorporation had been signed, Walcott found himself hired to investigate the geology of the Colorado plateau.

Although his position was initially temporary, such was his success that within a year it had been made permanent. He spent four years in the West before publishing his first major work, the *Palaeontology of the Eureka District*, in 1884. Confident that he was now safely embarked on a career in palaeontology, he returned East to work on his beloved Lower Palaeozoic rocks of New York state and New England. In 1888, thirteen years after the death of his first wife, he remarried. During a working honeymoon in Newfoundland he investigated the sequence of Cambrian rocks there and documented the perplexing similarity of the succession of trilobite faunas to those found in the Baltic states of Eurasia. (It would be more than sixty years until the theory of plate tectonics would explain that Newfoundland and the Scandinavian sequences had once been contiguous faunal provinces.) The newly-wed Walcotts travelled together to London to the International Congress of Geologists where, in a state of high excitement, Charles reported this new finding. During 1889, Charles plunged more deeply still into his work on the Cambrian, work that twenty years later was to lead to one of the most important fossil finds in the history of palaeontology. He took time out in 1889 to visit Toronto and attend the first meeting of the newly incorporated Geological Society of America, now recognised as one of the premier geoscience societies in the world. Charles Walcott was on a roll. Not only did he present a paper at the meeting, he was not reticent in commenting on the presentations of his peers. In the early 1890s Walcott travelled and published extensively on Cambrian and Precambrian rocks. So assiduously did he cultivate the American geological community and so high-profile was his work on Lower Palaeozoic fossils that in 1894 he was invited to succeed John Wesley Powell as the director of the US Geological Survey. With this his career truly took off and Charles Doolittle Walcott became the most influential geologist in America. A far cry from the young man who had started as a grocery clerk and who had sold fossils to make ends meet.

The rest of the decade passed in a flurry of administrative work. It became his practice to travel extensively in the summer – always with his family in tow – to escape the pressures of life in the capital and to retain contact with his palaeontological roots. By the turn of the century, Walcott was the indispensable first call for anyone who wanted advancement in the American geological scene, and a scientific 'fixer' of gargantuan proportions. At the end of 1901, Walcott met Andrew Carnegie, the philanthropist multimillionaire who had made his money from the steel industry. Within a matter of days, Walcott had finalised the details of how

Carnegie's gift could be best used and had laid the administrative foundations for the Carnegie Institution. Walcott worked exceptionally hard to get a reluctant Carnegie to bankroll a dedicated laboratory for the geosciences but in 1904 he succeeded and the Geophysical Laboratory was born, the same lab which five decades later Philip Abelson would take to such great heights.

In 1907 Walcott was offered and accepted the most prestigious job in American science, the Secretaryship of the Smithsonian Institution. At that time he was ready to resign from the Geological Survey, having been in the job for thirteen years, and was also happy to take a back seat in the administration of the Carnegie Institution now that he was sure that it was securely underway. Despite his new position, Walcott did not change the habit of half a lifetime and continued to geologise in the West during the summer with his wife, their two sons, Stuart and Sidney, and a daughter, Helen, to whom he was devoted. During his time at the Survey his interest in Lower Palaeozoic rocks and fossils had taken him to Idaho and Montana and, since geology does not kowtow to mankind's artificial frontiers, it was not long before his collecting took him northwards into Canada and the Lower Palaeozoic rocks of British Columbia and Alberta. The rest, as we have seen, is history: only two years after being appointed Secretary of the Smithsonian Walcott had discovered the Burgess Shale.

Walcott collected over 80,000 specimens in the subsequent years he spent visiting Burgess Pass. His approach was primarily to name his discoveries and four such papers appeared between 1911 and 1912. Walcott died on 9 February 1927, three years before Percy Raymond, Professor of Palaeontology at Harvard, went back to Burgess Pass and reopened the quarry that Walcott had abandoned in 1917. Raymond excavated a new quarry nearby but found only a few new species. Walcott's collections had been so comprehensive that there was little new to be found and so there the matter rested; most of the Burgess species were named and catalogued but their true significance was not investigated until the 1960s. Although these fossils were viewed as something extraordinary because of their exquisite preservation and antiquity little was done to integrate them into the rest of palaeontology. Darwin's original notion, that the Precambrian was devoid of life, was still dominant and so the idea that complex forms could already have evolved by the beginning of the Cambrian was largely ignored. Even in Walcott's day the fauna of the Burgess Shale was 'reliably' dated as being of middle Cambrian age. Everyone was quite sure that there could be no Precambrian fossils. Until, that is, a young Australian found the oldest fossils in the world outside an abandoned mine entrance in the Flinders Range of South Australia.

The X-files of evolution

To understand the significance of the Cambrian explosion it must be placed in context and that context is one of time – the large amounts of the Phanerozoic aeon and the depthless tracts of the Proterozoic and Archaean aeons which preceded it. We have a fair idea of what came after because we have, after all, the faunas of the Silurian world as seen through the eyes of Murchison, Lapworth, Elles and Wood: floating graptolite colonies, huge reefs bordering silent continents, and the very beginnings of land colonisation by the primitive precursors of rooted plants. The animals of the Silurian world – despite being separated from it by 120 million years (or almost twice the duration of the Cenozoic) can be traced directly back to many of the animals of Burgess Shale age; they are a scenario that we are familiar with.

But the Burgess animals are not the first animals that colonised this planet. These Cambrian faunas were themselves preceded by a strange and wonderful fauna that left only impressions of themselves in the rocks of England, Siberia and Australia. They take their name from the fossil locality in Australia – Ediacara – where they were first discovered in 1946 by one Reginald Sprigg. Sprigg was a fascinating character and an exceptional geologist. Born in 1919, at the age of seventeen he was elected the youngest ever Fellow of the Royal Society of South Australia, six years *before* he received his masters degree from the University of Adelaide and ten years before he made the discovery that would ultimately change the face of palaeontology. In 1946, Sprigg, then an assistant geologist to the government of South Australia, was examining some old silver and lead mines in the Ediacaran Hills nearly 400 miles north of Adelaide. The Ediacaran hills look like one of the sets from those old Mad Max films; they are ancient, heavily eroded hummocks in the desert surrounded by places whose names tell you all you need to know about the area: Hell's Gate, Devil's Peak, and the wonderfully named Reaphook Hill. In the austral summer the area can get so hot that it is impossible to pick up rocks. The only people who are at home in Ediacara are the aboriginals from whom the area takes its name: in their ancient tongue it means 'veinlike spring of water', which presumably indicates wishful thinking.

The commonly told legend has it that Sprigg's discovery was an accident; as he approached the old mines, his geologist's eye was pursuing its customary roving course even across the unpromising surface of the local outcrop, a formation known as the Pound Quartzite. The area had been mapped many years before by his old professor, the eminent Professor Sir Douglas Mawson of the University of Adelaide. Mawson was a geologist adventurer of the old school – he had been on Shackleton's 1907 expedition to Antarctica – and was an inspiration to the young Sprigg. It had been Mawson who had discovered radioactive minerals in the Flinders Range while mapping its late Proterozoic and early Cambrian rocks. Given their age – so the legend has it – the last thing that Sprigg would have been expecting to find was fossils. But, as he walked, Sprigg noted numerous impressions of what he took to be jellyfish outcropping in the Pound Quartzite. He collected them, took them back to Adelaide and wrote them up. Sprigg, who lived at a time when Precambrian life was thought to be a contradiction in terms, thought that the jellyfish horizon was of Cambrian age and did not appreciate that he had just broken the Precambrian life barrier. Thereafter, perhaps out of misguided loyalty to his *alma mater*, he published his descriptions in the obscure *Proceedings of the Royal Society of South Australia* where they languished for the next ten years until the brilliant European savant Martin Glaessner rescued them from obscurity.

The true story is not so simple. Reg Sprigg first met Glaessner in 1942 – right in the middle of the war and four years before he discovered the fossils at Ediacara – when Glaessner was employed by the Australasian Petroleum Company. Reg Sprigg was then an assistant research officer with the Commonwealth Scientific and Research Organisation based in Adelaide where he was working as a soil scientist.

Martin F. Glaessner was a remarkable man. Born in Bohemia in 1906, he had been made a research associate at the age of sixteen at the University of Vienna. In 1932, he was invited by the Director of the State Petroleum Research Institute of the USSR to organise research work in micropalaeontology for the purpose of correlating the zones and strata of the Soviet Union's oil-fields. Martin Glaessner was the man who established the palaeontological infrastructure of Russia's oil interests through his preoccupation with all things micropalaeontological and particularly the use of the planktonic foraminifera as a correlation tool. In 1934 he was invited to found a micropalaeontological lab in Russia. Thus he planted the seeds that were later exploited so successfully by Subbotina, Morozova and Alimarina, the female architects of the palaeontological revolution who kept the USSR's petroleum industry alive through the dark days of war and Stalin's purges.

In 1938, rumour of Glaessner's successes in the East had penetrated as far as the headquarters of the Anglo-Iranian Oil Company (better known these days as BP) and they ruthlessly recruited him. He was working on one of the first textbooks of industrial micropalaeontology, a project that Anglo-Iranian enthusiastically endorsed, but they also sent him and his new wife, the Russian ballerina Tina Tupikina, south, charging him with the task of setting up a micropalaeo laboratory for the company's subsidiary (the Australasian Petroleum Company) at Port Moresby in Papua New Guinea. When he met Reg Sprigg he was filling in time before returning to Papua New Guinea by putting the finishing touches to his textbook, *The Principles of Micropalaeontology*. (It is an important work, sure enough, but within the trade just about as far from Precambrian palaeobiology as it is possible to get and still call yourself a palaeontologist.)

Although the common myth has it that Sprigg stumbled on the Ediacaran fauna by accident, the fact of the matter is that when he met Glaessner he had for several years been busy mapping, in his own time, the Cambrian and older outcrops of the Adelaide area in general and the northern Yorke Peninsula in particular in the hope of finding traces of

Precambrian life. He had already discovered what he considered to be a primitive eurypterid (horseshoe crab) which he showed to Mawson. The Professor was unimpressed and proceeded to bin Reg's prize specimen. This so enraged Sprigg that he widened his search into the poorly known upper Precambrian and basal Cambrian of the Flinders Range. In 1944 he was appointed as an Assistant Government Geologist in South Australia, a job that entailed examining abandoned mine workings in the Flinders Range with a view to reopening them if possible. It was a perfect combination for Sprigg for he could combine his professional and palaeontological interests. It was this happy conflation of travel, time and ten years of diligent experience that led to his discovery of the Ediacara fauna in 1946.

Sprigg knew perfectly well the potential importance of the fossils and prepared a manuscript for *Nature*. It was rejected. In 1947 Sprigg tried again, reading a paper describing *Edicaria flindersii* before a meeting in Adelaide of the Australian and New Zealand Association for the Advancement of Science. Both Glaessner and Mawson were at the meeting and neither was impressed. They agreed that Sprigg's fossils were nothing more than fortuitous inorganic markings. A similar reaction at the eighteenth International Geological Congress in London in 1948 was the last straw for Reg Sprigg. His general depression at his inability to persuade people of the importance of his discovery, coupled with other demands on his time (by now he was on the staff of the Department of Mines at Adelaide University) diverted him from any further study of the Ediacaran fauna. But just as Sprigg's interest was waning, Glaessner's was growing. Mawson appointed Glaessner to the new post of senior lecturer in palaeontology at the University of Adelaide and pretty soon Glaessner was after Reg Sprigg's fossils.

In 1954, after eight years during which he had tried to convince an uncaring world of the importance of the Ediacaran fauna and just after he had assisted in opening up the politically important Radium Hill uranium mine, Sprigg resigned from the University. He realised that in all the years of fruitless persuasion and constant travelling he had neglected his family. It was time to make amends and to make some money. It was time to leave the field to Martin Glaessner. He set up his own company, 'Geosurveys of

Australia', and took on consultancy work for oil companies. He was so successful that in 1962 he set up his own petroleum company, Beach Petroleum, which by the late sixties had become an international player and was ultimately bought out. Reg Sprigg had made his fortune. After that, his interests turned to conservation and the need to preserve the outback. In 1968, he bought a derelict tract of land named Arkaroola in the north of his beloved Flinders Range which he proceeded to convert into a wildlife sanctuary and wilderness reserve. He died on his estate in 1994.

Before Sprigg left Adelaide, though, he gave Glaessner permission to study his Ediacaran fossils. The best specimens he had already deposited in the Tate Museum in Adelaide but the second-class materials were in Sprigg's own company vaults. To his credit, Glaessner never forgot Sprigg's generosity and named several of the most outstanding forms in Sprigg's honour. One in particular Sprigg appreciated – *Mawsonites spriggi* – a jellyfish-type fossil that linked Sprigg's name with that of his illustrious professor. By the early 1950s Glaessner – having acquired that robust Australian humour – named another specimen after the fauna's discoverer: *Spriggina floundersi*. 'Named in honour of Reg Sprigg,' said Glaessner, 'this is probably the lowest worm that ever lived.' Sprigg, a native Australian, retorted instantly that since the worm had a clearly defined head-end it was probably the most advanced animal of its time. After all the vicissitudes of Reg's professional life, and his loss of the credit that he must have thought rightfully his, one can only wonder what the real subtext of this exchange was . . .

Glaessner's investigation of the Ediacaran fauna marked a major re-invention of himself. All his working life he had been a professional, industry-orientated micropalaeontologist. But, by the late 1950s, his interests in Precambrian life had crystallised and he became one of the earliest practitioners of the palaeontological subdiscipline known as palaeobiology.

The late fifties were a time of revolution in the study of what were now quite clearly Precambrian fossils and Glaessner was there to exploit it. There were two revolutions, 12,000 miles apart, and Glaessner as the expert on Precambrian life was in a position to be a part of both. The first was on his back doorstep. Just after Christmas 1956, at the height of the austral summer, Hans Minchum, a schoolteacher, was walking with his friend Ben Flounders through the blighted place that is the Ediacaran hills. They were searching for more specimens of the Ediacaran fossils. Flounders was a keen mineralogist and an excellent photographer and they had every expectation of bringing back some good specimens or at least of getting some good shots of those that were too large to move. But they ran into a heatwave; in South Australia, at the height of summer, a heatwave is the same magnitude of problem that a snowstorm is in central Canada in December. Minchum and Flounders were unable to collect much but what they did find stimulated a return visit. They came back in spring 1957 (September) and collected enough material to make a significant presentation to the South Australian Museum. Glaessner, now established as reader in palaeontology at the University of Adelaide, naturally heard about these additional finds. In March 1958 the South Australian Museum dispatched an expedition, led by Brian Daly, the curator of palaeontology, for the sole purpose of collecting material from the Ediacara range. The six-man expedition, including two amateur collectors and one of Glaessner's graduate students, collected enough material in four days to fill two small trucks and a trailer. In May of the same year the area was proclaimed a fossil reserve under the control of the State Minister of Education and the South Australian Museum.

At about the same time, news began to filter through from Europe of an extraordinary find in the most ordinary of places: Leicestershire in the English Midlands. A young man, a schoolboy really, was walking through his native Leicestershire in Charnwood Forest, a woody refuge of middle England, when he came across a large fernlike impression embedded in the sandstone. John Mason did not know that the palaeontological creed dictated that fossils from before the Burgess Shale were not allowed, nor did he know of poor Reg Sprigg's broken heart in South Australia. Ignorance was bliss and he contacted a palaeontologist at the University of Leicester (a breeding-ground for many excellent practitioners since) named Trevor Ford who listened attentively to him and then returned with him to the quarry. Over the next few months they discovered several other fossils. *Charnia* and *Charniadiscus* were formally inducted into the palaeontological literature in 1958, the same year that the South Australian scientific community finally woke up to the wonders on their doorstep.

In October 1958 the first joint expedition of the South Australian Museum and the Department of Geology at the University of Adelaide was carried out; thereafter, systematic exploration at Ediacara continued for a number of years, resulting in the collection of over 1500 specimens. Other localities were also discovered from the same horizon – the Pound Quartzite – in other areas of the Flinders Range. Two decades later, in 1984, Glaessner, now retired, published *The Dawn of Animal Life* which is still one of the most influential books in the entire discipline of palaeobiology.

But what of these strange animals that were found in the ancient rocks of the South Australian desert in the dog days following the end of the Second World War? What do we know of their biology and relation-ships to other organisms? Many of them resemble jellyfish and to Glaessner this made perfect sense. Jellyfish and their relatives (the Ctenophores) are diploblastic – they have two layers of body tissue separated by an undifferentiated jelly called the mesoglea. All other animals are triplo-blastic; with three well-developed tissue layers. So modern-day jellyfish are primitive compared to the rest of the animal kingdom – they are atavistic throwbacks to a simpler level of cellular organisation – but the Proterozoic was so very long ago that it was easy from Glaessner to conclude that these jellyfish were the kings of the Precambrian world, from before evolution had had a chance to produce the revised, superior design of the triploblastic animals. The late Precambrian ocean must have been dominated by both free-floating and sedentary jellyfish designs. But other animals besides jellyfish are known from the Ediacaran-age faunas and some of these appear to have no living relatives. *Spriggina* itself, over which Glaessner and Sprigg had their little 'joke', may be a primitive arthropod but there is as yet no consensus.

Dickinsonia, an organism that looks for all the world like a pitta bread divided into semi-radial compartments quite certainly has no living relative, yet like *Spriggina* it was symmetrical about its long axis. Animals like *Dickinsonia* appear to have been entirely sedentary, living without movement on the sea bed probably in symbiosis with photosynthetic algae. They formed according to a construction blueprint that is not used at all in the oceans of today. Yet Glaessner tried to fit all members of the Ediacaran fauna into conventional pigeonholes. It was only as late as 1983 that Adolf Seilacher of the University of Tubingen came up with a different and radical notion which he called the 'Vendobionta'. Seilacher noted that the animals of the Ediacaran age do not fit into the same basic body-plan categories as the animals that are around today (or indeed that we think of as the Burgess Shale fauna). He noted that there were continuous gradations in form between animals that, if they were alive today, we would place in separate taxonomic groups. For example, *Spriggina* (thought by many to be a primitive arthropod) shares certain features with the Ediacaran sea-pen *Charniadiscus* found in Leicestershire. Seilacher now views the Ediacaran fauna as a group of wholly extinct animals, multicellular but constructed around a different type of tissue organisation from that of the rest of the known animal world. In his view, they were constructed of segments in a manner superficially like a quilted air-bed, and, as they lacked both head and gut, were a failed experiment in the history of life, belonging in the X-files of evolution. They were not ancestral to anything.

A more modern view of the Ediacarans reconciles both Glaessner's and Seilacher's interpretations. Recent discoveries of trace-fossils (marks made by the movement or presence of a fossil) suggest that the ancestors of the molluscs (animals well known from the Cambrian as well as the present day) were probably a part of the Ediacaran fauna and, along with several other groups, can after all be fitted into modern biological categories. But there are others such as *Dickinsonia* that have no representatives in the subsequent fossil record; their outré Precambrian designs were ruthlessly culled. So what happened to the 'air-beds' and the other strange represent-atives of the late Precambrian? There is growing evidence that there was a mass extinction (yes, yet another 'terminator') at the Precambrian–Cambrian boundary although there is as yet no consensus, and there are some who say that the Ediacarans do trickle on into the Phanerozoic. One thing though *is* certain, *most* of the Ediacarans did not make it beyond the end of the Proterozoic: they were creatures of a narrow slice of time, sandwiched between the world of bacteria and the world of the better-engineered triploblasts. The Ediacaran world lasted only 22 million years, the current best estimate puts their tenure at between 565 to 543 million years before present; 22 million years in which evolution engaged in a massive and perhaps unprecedented experiment in order to find out what would and would not work. Although some of the X-file Ediacarans may have stumbled into the Phanerozoic, their day was gone. Ancestors of the forms that we are familiar with today were the new kings of creation. All of which brings us back to the Burgess Shale.

It's life Jim, but not as we know it . . .

Today, one of the biggest questions that exercises the new palaeontologists concerns the origins of the complex animals, the Metazoa, whose cells are organised into tissues and complex body plans. The Metazoa are many-celled creatures, as opposed to the basic, single-celled Protozoa, and include everything that we are accustomed to consider an animal: the chordates (which we belong to), the arthropods (that include the insects), the frighteningly abundant nematodes (parasitic worms), the molluscs, the annelids (segmented worms), and the brachiopods, those superficially clam-like creatures restricted today to only a handful of species but so fecund in the Palaeozoic that they were a major rock-forming component of the Silurian reefs. All of these body types evolved apparently simultaneously about 540 million years ago, perhaps even contemporaneously with the doomed Ediacarans. The suddenness with which they appeared in the fossil record is so extreme that this episode is known as the 'Cambrian explosion'.

For thirty years the animals of the Burgess Shale remained almost unstudied. Walcott, constrained by his huge administrative burden, had managed to classify some – but only some – of the Shale's fauna into conventional groups. But as others (notably Stephen Jay Gould in his seminal book *Wonderful Life*) have observed, there was no hint in Walcott's writings that the Burgess Shale might represent something more spectacular. The start of the new phase of work on its fauna began in earnest in the mid-1960s (despite an abortive attempt by a lone Italian at the end of the 1950s to re-ignite interest). The impetus for the new study of the Burgess fossils was the fact that the Canadian Geological Survey had decided to remap the southern Rocky Mountains of Alberta and British Columbia. Harry Whittington, chief invertebrate palaeontologist at Harvard was co-opted on to the team studying the Burgess Shale because he was the world's greatest living expert on fossil arthropods. The head of the Canadian Geological Survey at that time was none other than Digby McLaren, whose ideas of an asteroid impact as a cause of mass extinction predated those of the Alvarez coterie by at least a decade. McLaren was keen that the Canadians should have some of their most spectacular fossil resource in their own museums, for both the previous expeditions, Walcott's and Raymond's, had taken their booty across the border to Canada's wealthy and influential neighbour. The expedition – originally planned to start in the field season of 1965 – eventually got to the target locality in the summer of 1966. You can imagine their excitement as they made their way up the deeply forested tracks from the small Canadian Pacific Railroad town of Field for the first time in thirty years to the rocky pass between Mount Stephen and Mount Wapta. They found both Walcott's and Raymond's quarries virtually unchanged. Especially poignantly, Whittington's expedition set out to find Walcott's original campsite and eventually located it below the treeline, in a dim and forbidding forest clearing, hardly altered by the slow-growing vegetation of the alpine Canadian wilderness. At places around the small clearing they could still see piles of carefully arranged bundles of wood, these were not preparations for unused camp fires, but the makeshift beds that Walcott, his wife, sons and assistants had slept upon for weeks at a time during their expeditions over forty years before. There were still raw patches in the grass where Walcott's tents had stood, in the trees a rusting meat-hook still hung. Among these pioneering reminders, the Whittington expedition worked with the advantages of modern technology, and helicopters carried supplies in and fossils out.

In 1966, Harry Whittington, a native Brit, knew that he was going to be returning to Britain, as he had been appointed to the Woodwardian Chair of Geology at the University of Cambridge. This was the time when Harry Godwin held a parallel Chair in the Department of Botany and through his far-sighted vision was installing the first electron microscopes and thinking about hiring someone (Nick Shackleton as it turned out) to weigh the world's ice sheets. There are two connections here: the Woodwardian Chair had been held six decades before by Johnny Marr – the original mentor of Elles and Wood who had read their seminal papers on the Wenlock and Ludlow periods before the Geological Society of London – and, in the nicest symmetry of all, Walcott had named the most common of the Burgess arthropods *Marella* after that celebrated Cambridge palaeontologist. And so, by coincidence, the resurvey of the Burgess Shale relocated to England, to Cambridge. Two other researchers were recruited by Whittington to help with the re-study of the Burgess Shale: David Bruton and Christopher Hughes each took one of the Burgess Shale taxa as their personal project. Bruton took *Sidneyia* (named after Walcott's son who discovered the first specimen) and Hughes took *Burgessia*. Whittington gave the first paper from this new wave of study on the most common of the Burgess arthropods (*Marella*) before a meeting in Chicago in 1969. Yet it was already becoming clear that the Burgess Shale would require a larger research effort. Some research students were needed to do the legwork: Derek Briggs, an Irishman who approached Whittington and was given leave to work on the arthropods, and Simon Conway Morris.

Conway Morris became interested in the Burgess Shale in the early 1970s when, as an undergraduate at the University of Bristol, he was studying under the great Bob Savage. In common with many palaeontological museums across the world, Bristol had a small collection of Burgess fossils, teasers that had been sent out by Walcott, and a few specimens had been laid out for a practical lesson in the palaeontology course. It was enough to ignite Conway Morris's interest. After writing to Whittington, Conway Morris was invited for an interview and since Briggs had already claimed the arthropods he accepted Whittington's offer of the worms as the subject of his PhD studies. He started work, as is customary with British research students, in late September. The year was 1972. Almost immediately Conway Morris's researches were interrupted by appendicitis but, notwithstanding that, by March 1973 he and Briggs had examined all the material collected by the Canadian Geological Survey's expeditions of the 1960s. The techniques that they used were photography, particularly under UV light, as well as detailed dissection of the fossils and laborious drawings made using a *camera lucida* (essentially the same technique used by Lapworth, Elles and Wood for drawing graptolites). Such work is extremely painstaking but there is one set of volumes that made extensive use of *camera lucida* drawings. When I entered palaeontology in the early eighties the subject was dominated by them: a huge set of blue-bound books known as the *Treatise of Invertebrate Palaeontology*. The set consisted of twenty to thirty volumes, each dedicated to a different group of fossil animals. It was the ultimate monument to the old science of fossils – and the dusty attitudes of the curator at the end of *Raiders of the Lost Ark* – for all of palaeontology was then considered to be summarised in these volumes. By cataloguing the dead, these volumes somehow professed to contribute to a living science.

These were the books that dominated the bookshelves of Bristol's palaeontology laboratory. The animals of the Burgess Shale that were catalogued, and written up by the Swedish scientist Leif Stormer in 1959, were included in the fat tome that dealt with the trilobites. Stormer had disagreed with Walcott's original approach – which was to fit, or shoehorn (to use Steve Gould's term), the Burgess Shale fauna into existing groups of arthropods. Instead, Stormer put almost all of the Burgess forms into a catch-all sister group of the trilobites proper which he called the *Trilobitoidea* (literally 'trilobite-like'). Stormer then put the *Trilobita* and the *Trilobitoidea* into a larger taxonomic grouping – the *Trilobitamorpha*. But Stormer had to concede that some of the Burgess fossils were too strange to fit even into his inelegant grab-bag of a taxon. These he put in their own catch-all category, starkly named sub-class 'uncertain'. By the early seventies Whittington had proved that *Marella* – Walcott's 'phyllopod' – bore little resemblance to trilobites proper. By 1975, Whittington had shown that *Opabinia* – one of the strangest Burgess animals with a long flexible proboscis that looks for all the world like the hose of a vacuum cleaner with five eyes – was not an arthropod either. It was *Opabinia* that proved to Whittington beyond doubt that at least some of the Burgess animals, like the Ediacarans before them, could not be accommodated in existing groups. While this revelation was occurring to Whittington, Conway Morris was embarking on a series of five papers in which he described some of the strangest of the Burgess animals and with which he established his career. These papers, published in 1976 and 1977, were all short and therefore quite different from the mainstream of palaeontological publications (typically the monograph was little changed in style since the days of Elles and Wood). Conway Morris's researches during those few years persuaded him that the bulk of the Burgess Shale fossils could not be accommodated in known groups of animals. Indeed *Hallucigenia* with its prolonged legs stands as the ultimate monument to the strangeness of the Burgess Shale fauna.

It was this conclusion that stimulated Stephen Jay Gould to immortalise Whittington, Briggs and Conway Morris in *Wonderful Life*. Gould took the name for his book from the 1940s' James Stewart movie in which a young man seeks to leave a home in small-town USA but through a combination of contrivances is prevented from doing so and gradually comes to appreciate how important his 'ordinary' life is. As always with Steve Gould's writings the deft and humorous touch implied by the title and his peerless prose conceal a heavyweight scientific message. In this case a particular, favourite theme of his, 'contingency'; blind chance has a role to play in evolution. The idea that the vagaries of fate have had a hand in the fortunes of life on Earth is a thread that runs through much of Gould's later writings, particularly in his celebrated series of essays published in *Natural History* magazine. In *Wonderful Life* Gould reiterates these, and then combines them with the central idea that the Cambridge three had eventually to come to terms with; namely that the animals of the Burgess Shale represent a broader spectrum of structural blueprints (or *bauplans*) than are found today. To Gould, the history of life is a pyramid with most of the original experiments in evolution removed before they had a chance to get going; the wheel of fortune which selects the lucky winners in life's lottery is spun by the blind hand of chance. This view of life is quite different from the classical view of evolution (as promoted by the Trinity of Simpson, Mayr and Dobzhansky) that life tends to increase in both diversity and complexity through time by the infinitesimal accumulation of favourable mutations over aeons. Gould's book is a masterpiece, scientifically and historically accurate, and brimming with his customary literary chutzpah.

The central idea of *Wonderful Life* (that an experimental suite of designs was winnowed by the whim of chance operating on the floor of an ancient ocean) has recently been challenged by none other than Conway Morris himself in his book *The Crucible of Creation*. It seems that Conway Morris disagrees with Gould's thesis of the importance of chance in evolution, and remarkably berates Gould for the thesis advanced in *Wonderful Life*. It would seem that Conway Morris has become an adherent of the older view, that life tends to greater complexity over the course of geological time.

There is now a growing understanding of the true nature of the Cambrian explosion. Since Gould's book two other important faunas have been the subject of intensive study: the Sirius Passet fauna and the Chengjiang fauna. The Sirius Passet fauna in Greenland is of Lower Cambrian age, significantly closer to the Precambrian–Cambrian (or Proterozoic–Phanerozoic) boundary than the Burgess Shale fauna itself. Like the Burgess Shale the Sirius Passet fauna also comes from Laurentia, the continent formed by present-day Greenland and North America which lay across the palaeo-equator in the centre of a giant ocean. And also like the Burgess Shale, the Sirius Passet fauna was deposited in deep waters – probably by a mudslide – in similar fashion to the Burgess animals. Although there are many arthropods in the Sirius Passet there is only one species of trilobite, in contrast to the Burgess Shale's dozen. The Sirius Passet fauna also has very few fossils with shells, perhaps because the fauna was preserved in an area with very low oxygen and selective preservation of fossils with hard parts did not occur.

The Chengjiang fauna is almost exactly the same age as the Sirius Passet, but importantly, this fossil locality was not from Laurentia but from another palaeocontinent completely, the South China craton several thousand miles away. The Chengjiang fauna is exposed on hillsides about 30 miles south-east of Kunming in Yunnan province, China. This fossil fauna has a history almost as exotic as that of the Burgess Shale. It was found in the first decade of the twentieth century by the French Geological Survey working in Indochina and may even predate Walcott's discovery of the Burgess Shale. However, the initial reports were published in obscure journals and it was not until 1984 that the scientific study of Chengjiang started when Chinese palaeontologist Hou Xianguang located the principal fossil-bearing beds. The Chengjiang fossils are spectacularly well-preserved as red-brown impressions in a yellow shale. There are strong faunal similarities to the Burgess Shale fossils despite the fact that Chengjiang is not only older but was deposited on a separate continent. There are several arthropods with well preserved appendages and also examples of other Burgess Shale groups. Curiously enough it was one of the Chengjiang arthropods which suggested to Hou Xianguang and Lars Ramskold that Conway Morris's original reconstruction of the very strange organism *Hallucigenia* was incorrect. It was Ramskold who dissected one of the specimens described by Conway Morris and showed that one of the flexible appendages was actually tipped by a claw, proof positive that it was in fact a leg. Conway Morris had reconstructed the animal upside down.

The most exciting of the recent discoveries at Chengjiang is of an animal that appears to be a close relative of the Burgess Shale's *Pikaia*, a chordate and therefore close to the evolutionary line that gave rise to the vertebrates and ultimately us. The Chinese fossil is called *Cathaymyrus*, literally 'Chinese eel'. Another strange fossil from Chengjiang is *Yunnanozoon* which is thought to occupy a primitive position close to Elles and Woods' beloved graptolites. Taken together it is clear that the Sirius Passet, Chengjiang and Burgess Shale have much in common. The Burgess Shale fossils are not unique after all. As recent discoveries have accumulated, the need to see the Cambrian as something unusual and extreme has now receded. This is well-symbolised by the fact that *Hallucigenia* – strangest of all the Burgess animals – is now comfortably accommodated in an existing group. These lobopods (as they are informally known) are a group with only a couple of species left in it, the best known of which is *Peripatus*, a land-dwelling, worm-like creature with both arthropod and annelid characteristics that is to be found scuttling under logs in the jungles of the southern hemisphere. I can still remember *Peripatus* featuring in *Life on Earth* – David Attenborough's terrific TV series from the late seventies that first brought me into palaeontology – and being hailed as one of the few examples of a 'non-missing' link. *Hallucigenia* of the Burgess Shale was merely an early casualty of a group that still exists on Earth. Recent finds from Herefordshire by my colleagues Derek and David Siveter and Derek Briggs suggest that the lobopods were important in the Silurian too.

So the idea that the Cambrian was a time of evolutionary experimentation in which chance rolled the dice and removed swathes of body plans on a whim is no longer necessary. But it *was* necessary to have the debate. And for this we should all be grateful to Steve Gould.

But this new understanding of early Cambrian realities merely pushes the problem of how modern, complex life (the Metazoa) originated further back in time, either to the base of the Cambrian or beyond, into the last millennia of the Proterozoic. Had the innovations which appeared during the late Precambrian or earliest Cambrian been simmering in the biological cauldron of the Proterozoic for aeons or did life really arise rapidly? What has been termed the 'sudden arrival versus slow-burning fuse' debate turns on a disparity between the data of the fossil record and that from the reluctant palaeontologists.

The ghost in the machine

If the fossil record is really inadequate to answer the question of just when the metazoan body plan evolved then we must turn to some other form of investigation; this effectively means measuring the time of separation of different lineages as recorded in their nucleotide or amino-acid sequences; the technique pioneered by Zuckerkandl and Pauling (and discussed at length in Chapter 8). In 1996, Greg Wray and his colleagues Jeffrey Levinton and Leo Shapiro at the State University of New York attempted just such an investigation. They chose seven genes that are common to both protostomes and deuterostomes (the two great groups that together comprise the Metazoa) and which are essential to the functioning of cells: they are called 'housekeeping genes'. The genes that they used were not related to each other – some were from mitochondria and others were from cell nucleii and they manufactured very different proteins. All of this helped to cancel out variations in the speed of the different molecular clocks. The result of the Wray team's investigation was conclusive – all seven gene groups agreed that *all* the phyla examined had diverged between about 1.2 and 1 billion years ago – without exception within the Precambrian. The base of the Cambrian is about 540 million years ago. On the face of it then, these groups of animals started to diverge over twice as long ago as would have been thought from reading the fossil record. The Wray study also suggested that the split in one lineage (between the chordates and the echinoderms) occurred about 1 *billion* years ago, still way before the base of the Cambrian. This was the study that gave rise to what came to be known as the 'slow-burning fuse hypothesis', that modern animal groups attained a specific genetic identity way back in the Precambrian (indeed at the end of the Mesoproterozoic – the middle 'era' of *pre*-ancient life) but yet are not seen in the fossil record because they lacked the hard parts to leave as fossils. This was quite at variance with the Cambrian explosion hypothesis, namely that exceptionally high rates of evolution at the bottom of the Cambrian had given rise to more phyla than are represented in the modern-day range of body plans.

The Wray team pointed out that the Cambrian explosion idea implied rates of mutation an order of magnitude higher than those considered to be 'normal'. They also noted that, if their idea of very early divergence times was accurate, it would imply that even the strangest Ediacarans like *Dickinsonia* and *Spriggina* should probably be placed among existing groups (just as Martin Glaessner had originally predicted). The Wray group acknowledged that their divergence dates were not at all the same thing as dates from the fossil record. (The latter indicate when something has evolved and become sufficiently distinctive to leave a record in the rocks; the former indicate simply – and roughly – when genetic intermixing has ceased.) Still, the two estimates of the origination of animal taxa needed reconciling. Attempts made to find Proterozoic metazoan fossils met with some success. Duncan McIlroy and Martin Brasier of Oxford University have already found what they claim to be a trace fossil with an age of about 600 million years but even this is not enough; there is still a huge mismatch between the record of the rocks and record of the genes. Some suggested that the 'slow-burning fuse hypothesis' was correct and that the 'something' that happened near the base of the Cambrian was simply that fossils became larger and acquired hard parts (in the process becoming 'visible' to the fossil record). Yet it is important to remember that many small animals, no matter how numerous, are unknown as fossils. The late Precambrian, say the 'slow-burn' supporters, may have been teeming with these animals and yet they left no trace of themselves. But there are disagreements even within the reluctant palaeontologists.

In 1998, a group led by the noted American geneticist Francisco Ayala published a paper that suggested that the Wray team's study was fundamentally flawed. The Ayala team studied no fewer than *eighteen* genes (including six of the seven originally mapped by the Wray group) and yet reached radically different conclusions. They found that the protostomes diverged from the deuterostomes only 670 million years ago (as opposed to the Wray group's estimate of 1.2 billion years ago) and that the chordates diverged from the echinoderms only 600 million years ago; an age that places them squarely around the time of the Ediacarans. Both of these estimates agree well with the data from the fossil record. The Ayala group were not hesitant in putting their finger on the error that the Wray group had made. Their clock, they said, simply ran too fast. Take for example, three travellers journeying from Los Angeles to San Diego, San Francisco and New York respectively. Each journey will necessarily show a strong correlation between distance and the time needed to arrive even though one is travelling by car, another by train and the other by plane. It takes time to get there, but one cannot make any assumptions about how much time is involved unless one has a clear idea of the vehicle used. It would be folly to average the time/distance rate between Los Angeles and the other three cities using different modes of transport and then use it to estimate the time needed to travel between, say Los Angeles and London. This, they said, was the mistake that the Wray group had made and which had enabled them to arrive at such ancient divergence dates.

As I write this, it would seem that the answers from the reluctant palaeontologists and classical palaeontology are converging. There is now good agreement between the estimates from the fossil record and the estimates from gene sequences that the protostomes and deuterostomes originated and diverged in the late Neoproterozoic, during the 160 million years that preceded the base of the Cambrian. In the first year of the twenty-first century there is already yet another sea-change in palaeontology. In order to find the first metazoan fossils, we are being guided to rocks of the right age by the techniques of molecular biology.

Frozen at the start of time

Although molecular biology may be creating the agenda for studying the dawn of animal life there are several things that molecular biology cannot illuminate, and these are the environmental forces that shaped the evolution of these strange and distant organisms. This question – one that is shorthanded as the 'extrinsic' factors that drove the Cambrian explosion – is being approached through a combination of isotope palaeontology and classical geology. The emerging story is that the latest Precambrian and earliest Cambrian were an extraordinary time indeed. There is evidence from the carbon isotope record of very large-scale oscillations in Earth's carbon cycle close to the Precambrian–Cambrian (or Proterozoic–Phanerozoic) boundary. These very large fluctuations in the carbon isotope record have a maximum extent of *14 parts per thousand*, which is about seven times as large as the carbon isotope change recorded at the K–T boundary. With one exception, the carbon isotope anomalies at the Precambrian–Cambrian boundary are larger than anything known from the rest of the Proterozoic and vastly larger than anything known from the Phanerozoic. The carbon isotope shifts of the Pleistocene – which as we have seen reflect the waxing and waning of atmospheric carbon dioxide as Earth cycled in and out of the most recent series of ice ages – are spasms of no consequence by comparison. The huge carbon isotope shifts of the latest Proterozoic imply large – no, huge – climatic shifts. The very positive carbon isotope values that precede the profound negative blips in the record suggest that the rate of oceanic production and carbon burial in the late Proterozoic must have been, at least episodically, very high indeed and the lows are so negative that they approach the carbon isotope ratio of volcanic emissions, the second most negative (the first being that of methane hydrates) known. These lows imply profound extinctions among latest Proterozoic life.

The high rates of carbon burial inferred for the late Neoproterozoic would have the effects of removing carbon dioxide from the atmosphere, preconditioning the world for cooling, the so-called 'ice-house' effect well-known from later parts of the geological column. The carbon isotope minima are associated with fossilised glacial debris suggesting that the extinctions were associated with glaciations. But here the story takes a strange and menacing twist, for it is known (on the basis of palaeomagnetic evidence) that Namibia, where many of these glacial rocks are found, straddled the equator during late Proterozoic times. Could it be that polar ice really extended as far as the equator? In fact, this is not a new idea. Brian Harland, Professor of Geology at Cambridge in the mid-1960s, suggested just such a scenario based on the evidence from the then new science of latitudinal reconstructions based on palaeomagnetic evidence. In collaboration with the brilliant Martin Rudwick, the maverick palaeontologist whose light touch was associated with many major palaeontological developments of the sixties and seventies (not least the idea of functional morphology), they speculated that a period of warm climate followed what they referred to as the 'great Neo-Proterozoic ice-age' and it is this that allowed life to evolve beyond the single-cell level. As Harland and Rudwick were developing these notions, the Soviet climatologist, Mikhail Budyko (then working in the fledgling field of modelling climates with computers), realised the importance of ice-albedo feedback in the control of ice ages. Ice-albedo feedback is the mechanism by which ice sheets expand. Since they are white and therefore reflective, as they grow they reflect progressively more of the sun's energy back into space. The bigger they are the more energy is turned away thereby allowing the ice sheets to grow still larger. Budyko's calculations revealed that if sheets of sea-ice encroached within

30° north or south of the equator the ice-albedo effect would be unstoppable, the positive feedback loop would continue until the whole globe was frozen over. In the sixties, the mechanism by which the ice-albedo feedback effect was prevented from going over the magic 30° dividing line was unknown. It was only later, in the seventies and eighties, that the importance of carbon dioxide came to be appreciated and it was realised that it was the complex interplay of albedo and CO_2 that kept Budyko's positive feedback loop in check.

But the idea of a global glaciation continued to exert a morbid fascination. After all, Harland and Rudwick's equatorial glacial debris-fields still needed an explanation. In 1992, the American palaeo-climatologist, Joe Kirschvink, pointed out that iron was commonly to be found mixed in with late Neoproterozoic glacial debris. Similar iron deposits were known from earlier in the Neoproterozoic where their laminated structure had led to the name 'Banded Iron Formations' (BIFs for short). Iron is almost completely insoluble in the presence of oxygen, but in the absence of oxygen it can dissolve and ultimately precipitate out as these rare formations of rock. The presence of BIFs from the earlier Proterozoic made sense, for this was the time before the first prokaryotic cells had started to flood Earth's primitive atmosphere with oxygen. But to find them at the end of the Proterozoic, right on the doorstep of the Phanerozoic? Kirschvink suggested that the answer might be related to the global glaciation proposed by Harland and Rudwick, a glaciation that Kirschvink provocatively labelled the 'snowball Earth'. An ice sheet that covered the world would, he suggested, eventually deprive the oceans of oxygen, and thereafter any iron that was introduced into them (for example through the exhalations of submarine vents at tectonic plate boundaries) would simply accumulate in the water. If oxygen were later to be introduced into this system then the iron would be forced to precipitate out – leaving behind Banded Iron Formations such as those found in Namibia.

Kirschvink also suggested that ice-meltback would eventually happen automatically as submarine and terrestrial volcanism continued to add CO_2 to the atmosphere. Eventually so much CO_2 would have been pumped into the atmosphere that the ice would start to retreat and the dissolved iron would automatically precipitate out.

The carbon isotope work and general geologising of two Harvard geologists, Paul Hoffman and Dan Schragg, have added several additional components to this rich apocalyptic mixture. First, the several oscillations in the carbon isotope record imply that there was *more than one* 'snowball Earth event', a conclusion supported by the known occurrence of more than one glacial deposit in the late Neoproterozoic. Second, their proposed controlling mechanism (a heavy carbon isotope record = enhanced CO_2 burial and therefore cooling; a light carbon isotope record = extinctions in the biosphere) was similar to the controls proposed for other glacial and extinction events in younger parts of the geological column. Third, the presence of warm shallow-water carbonate deposits immediately above the glacial debris lent support to Kirschvink's implication that the change from the snowball Earth to the super-greenhouse Earth would be very rapid indeed.

There were a few refinements added to the brew – Hoffman and Schragg's paper in 1998 calculated that the glacial episodes probably lasted several million years and the super-greenhouse episodes lasted only a few thousand; predictions in accord with Kirschvink's hypothesis. They speculated that the trigger for the onset of these extreme glaciations was the supposedly unique position of the continents in the late Neoproterozoic. After the break-up of the supercontinent Rodinia about 900 million years ago, large continental blocks were spread around the equator. This increased the global ratio of sea to land and resulted in increased rainfall which would have had the effect of 'scrubbing' carbon dioxide out of the atmosphere and preconditioning Earth for ice-house cooling. For life this was the era of boom-and-bust; the cycles of extreme cold and extreme heat would have induced a kind of environmental filter that repeatedly killed off the fledgling flora and fauna of primitive Earth. In a sense it was a return to the *Wonderful Life* scenario – life could have been very different, but every time it tried to get started a global glaciation came along and killed it off. Finally after the last snowball Earth event life *was* allowed to get on with it, and it was then that the Ediacarans evolved.

What can we make of this theory? On the one hand, it appears plausible. There is evidence for a cycle of glaciations which appeared to have spread as far as the equator in the late Proterozoic and may well have had an effect on the evolution of life, but if we look at the details of the Hoffman and Schragg theory we find some inconsistencies and problems. The most worrying of these is what preconditioned Earth for these big freezes in the first place. Hoffman and Schragg suggest that it was to do with the break-up of the Proterozoic supercontinent Rodinia that induced enhanced rainfall. How can such a hypothesis be tested? You may recall that a similar idea was floated by Simms and Ruffell for the end-Triassic (*see* Chapter 7) but their suggestion was backed up by the presence of a consistent fossil topography (karsts). Another crucial question is why has this not happened since? Hoffman and Schragg suggest that there is a 'carbon dioxide safety switch': as high-latitude continents get covered in ice, less carbon dioxide is buried. This remains in the atmosphere and prevents the crucial 30° north/south latitude from being reached. But the continents have been arranged along the equator at least once since the late Neoproterozoic, indeed in the early Palaeozoic, an era well known even to Elles and Wood, and yet we had no 'Snowball Earth Event' then. Finally, what was it that happened to the Ediacarans that allowed the *eleven* groups of modern Metazoa to flourish? The snowball earth hypothesis is silent on that point.

Final frontiers

We see clearly now that the new science of fossils is an amalgam of techniques and ideas culled from many other disciplines. As a science it has finally come of age and it is right that it should seek new horizons. The widest horizon of all opened up in 1996 when David McKay and his co-workers published a report in *Science* suggesting that a Martian meteorite from the Allan Hills of the Antarctic (ALH 84001) contained evidence of fossil life on Mars. This opened a can of worms of which the first was: how were the McKay team so sure that the meteorite that they measured was from Mars? The answer lay with a discovery made in the immediate aftermath of the Apollo missions that certain meteorites (particularly those found in Antarctica) had identical chemical characteristics to rocks brought back from the moon. This was the first inkling that meteorites found on Earth could be fragments of other bodies in the solar system. Following the Viking missions to Mars, which analysed some of the gases present on the planet using chromatographic techniques, it was found that some meteorites on Earth contained trapped gases which correlated perfectly with the gas composition of the red planet as known from the Viking missions. These meteorites were named SMC meteorites and they all had one common characteristic: they all had to be from Mars.

In 1984 an SMC meteorite was found that was catalogued as ALH 84001. This was different from other SMCs in that it contained fractures that were filled with a type of carbonate only known to form in enclosed depositional basins on Earth. These carbonates were important evidence that suggested that ALH 84001 must have been in proximity to water at some point in the geological past (water is considered to be the essential prerequisite for life based on the carbon atom). McKay and his team continued their researches and eventually published the *Science* paper that, on the basis of four lines of evidence, suggested that ALH 84001 showed evidence of fossil life on Mars. All four lines of evidence were necessarily circumstantial. The first was that the carbonates in ALH 84001 were formed at a temperature consistent with life, the second was that organic chemicals found in the meteorite were depleted in carbon-13 and therefore were traces of Martian biochemical reactions (PAHs), the third was that the magnetic grains in the rock came from bacteria (as they often do on Earth); and fourth – and most spectacular of all – that the strange structures found within the meteorite were the fossilised remains of bacteria. All four lines of evidence depend on life on Mars being organised along essentially the same lines that earthly life is, *but we do not know for sure that life based on other elements and biochemical systems could not exist.*

In the immediate aftermath of the paper, the McKay team came under intense criticism from the scientific community. Of the original claims made, it was the presence of the bacteria-shaped objects (BSOs) that they subsequently withdrew most notably from. Agreeing with others (like the noted Precambrian life specialist Bill Schopf) that most of the BSOs were indeed too small to have harboured the biochemical machinery necessary for life they subsequently found larger BSOs (up to 0.75 microns long) that *are* similar in size to earthly bacteria. New work has also found that structures thought to be the remains of Archaean bacteria are of similar size to the smaller BSOs originally noted by the McKay team. The possibility now exists again that these enigmatic structures are, after all, the direct fossil remains of Martian bacteria.

Although the McKay team have now conceded that the PAHs they found need not necessarily be evidence of life, it is clear that some of the organic material in ALH 84001 is likely to be of Martian origin. Since 1996, the team have continued their studies on other meteorites (the so-called 'Nakhla' and 'Shergotty' specimens particularly) as well as ALH 84001. They have since found that the magnetic particles within ALH 84001 are organised into chains and that these structures compare favourably with the organisation of magnetic particles by bacteria on Earth. These early bacteria left traces, known as biofilms, which have also been found in rocks of the Proterozoic aeon. McKay's team are currently claiming that they see evidence of biofilms in both ALH 84001 and the Nakhla meteorites.

These claims are controversial, not least because the McKay team continue to extrapolate directly from the characteristics of earthly life. This may not be a valid approach but, as they say, what other approach is there? Teams of scientists around the world are currently working together to try to establish objective criteria for the recognition of life on other planets using approaches based on the fundamental physical and chemical principles necessary for life to function (for example, the presence of the basic biochemistry of photosynthesis).

It is hoped that the establishment of such general principles will take the new science of fossils a stage further and establish the criteria needed to recognise life on other planets. In many ways, the approach is as fundamental as that taken by the early founders of the new science of fossils back in the heady years of the late nineteenth century. In the immediate future, two more spacecraft are scheduled to go to Mars to search for evidence of fossil life; the first NASA robotic return mission to

Mars is expected to bring back samples of the Martian surface by 2010 and the European Space Agency will soon be sending the 'Beagle 2' to Mars, commemorating in the process Charles Darwin's epochal voyage. It is due to land on Boxing Day 2003. NASA is now planning a manned mission to Mars in the second decade of the twenty-first century, and palaeontologists are already being consulted over the form of the mission.

The investigation of deepest Archaean time as well as planets beyond our own is the next challenge for the new science of fossils. What wonders await us now that we finally have the technology to ask the right questions? Will we find life in the universe beyond our own world? Will we find that life on our planet was originally seeded from space?

Only time, the ultimate essence of palaeontology, will tell . . .

The Holistic Palaeontology

If there is an encompassing theme to this book it is surely that the new palaeontology is unique among the sciences in its plurality. The new science of fossils operates by drawing together strands from many other disciplines. To select only a few examples from the preceding pages, we have seen how cladistic classification methods and the history of life as encapsulated in information-carrying molecules comes from biology; how our ability to date fossil-bearing rocks by isotopic techniques and palaeomagnetic changes comes from physics; and how our ability to trace the changing size of ice caps and understand extinctions in the biosphere comes from chemistry.

It turns out that the three sciences of my A-level youth were needed after all! And it turns out too that the new palaeontology – the schizophrenic science – is a thieving magpie that routinely raids the nests of other sciences to build its own. This is healthy; a new science is being born at the confluence of many others. This is the way that science should happen: where once there was nothing, just the transition from one major discipline to another, something new *should* grow.

The new science of fossils has traditionally been practised in university earth science or geology departments. It is here that specialists from within geology can interact and share ideas. But perhaps the new science of fossils requires a still wider remit? Perhaps the unique cross-disciplinary approach of the new palaeontology requires specialised research and teaching units that pool scientists from *all* the disciplines that contribute to the new science of fossils. One thing is sure, as we contemplate our uncertain future the only useful knowledge that we will garner will come from a study of our planet's past. But these 'holistic palaeontological research institutions' may be no more than a pipe-dream. There is currently a trend for universities and government agencies to atomise and reduce science to convenient funding-oriented units, a strategy that does not work well for a subject as broad as the new science of fossils. Once again a more flexible approach is needed to fund a subject that is more than the sum of its parts.

And finally, there is the matter of palaeontology's outdated image. We have seen that description-based palaeontology is only one facet of the much broader spectrum of the new science of fossils. It has its place, but it is now only a minor player in the much larger and more exciting endeavour that is a genuine interpretation-based science. Palaeontology is changing and it is right that it should. To think otherwise is to discredit the many who made this new discipline.

It is to dishonour the Architects of Eternity.

Glossary

Allopatry. Literally 'in another place'. Refers to the idea that small, isolated populations of a species are more likely to form new species than sub-sets of a bigger population because changes in the genetic makeup of the new species stand a better chance of being able to propagate in a smaller gene pool.

Amino acids. Organic molecules that are the fundamental building-blocks of proteins.

Ammonite. A loose term for a member of the mollusc sub-class *Ammonoidea*, four-gilled cephalopods related to the present day *Nautilus*. Their spectacular shells – often several tens of centimetres across – are widely found in marine Mesozoic rocks where they are widely used for correlation purposes. The ammonites were one of the major casualties of the K–T boundary extinctions.

Archaean. Oldest of the two eons that comprise Precambrian time – from about 3.9 billion years to 2.5 billion years ago. The Archaean is characterised by very rapid tectonic plate movements (driven by the huge energy of the still cooling Earth) but without large continents. By the Archaean-Proterozoic boundary though the planet was beginning to stabilise and plate movements more similar to those we are familiar with began to dominate.

Bathonian. A stage of the middle Jurassic (175–169 million years before present). Carbonate fossils and rocks are common components of these rocks where they outcrop in middle England. William Jocelyn Arkell studied actively in this area.

Benthic. Sea-floor dwelling.

Biostratigraphy. The discipline of recognising the passage of time in sedimentary rocks and of correlating them using the fossils that they contain.

Blue-green algae. Primitive form of single-celled life which derives its energy from photosynthesis. The cellular structure is prokaryotic.

Brachiopods. Bottom dwelling marine invertebrates with two symmetrical calcium carbonate (or protein) parts to their shells. Superficially they resemble certain molluscs but are in fact quite different. Also known as lamp-shells because of their resemblance to early Roman oil lamps they are now restricted and uncommon in the world's oceans. However, in the Palaeozoic they were one of the most abundant and diverse forms of marine life.

Calcite compensation depth. The depth in the ocean at which the rate of supply of carbonate (in the form of planktonic fossils) from surface waters is balanced by the rate at which they are dissolved by corrosive deeper waters.

Calcium carbonate (calcite). The mineral that makes chalk and limestones. It is commonly found in the geological record because since the Cambrian many animals have exploited its hardness to construct their shells. Examples of calcite secreting animals are found in most invertebrate groups and include the molluscs, the echinoderms, the brachiopods and the foraminifera.

Carbonate. See calcium carbonate.

Carnian. A late Triassic stage between the Landinian and Norian stages.

Cerium/lanthanum ratio. A ratio of rare-earth elements that is a useful measure of the intensity of oxygen deficiency in ancient oceans.

Chalk. A whitish, crumbly limestone made up predominantly of the remains of foraminifera and coccoliths which is particularly common in the south of Britain and north-western France where it gives its name to the geological formation known as the Chalk (note capitalisation).

Chloroplast. The cellular organelle that houses the photosynthetic machinery of green plants.

Chooze. The chalk–ooze transition. Commonly found at variable depth in deep ocean oozes where pressure and temperature have become sufficiently great to start compacting ooze into chalk.

Chordates. The group of animals that possess a rod of flexible tissue at some stage in their life cycle. In the sub-group known as the vertebrates the notochord is protected inside a bony, articulated column, the backbone.

Chromatography. A technique for separating molecules from more complicated mixtures which relies on the differing rates of diffusion of molecules of different size.

Cladistics. Classification scheme devised by the German entomologist Willi Hennig and published in his book *Phylogenetic Systematics* in 1950. Emphasises the possession of shared, derived characters to relate organisms without reference to any supposed evolutionary history.

Coccolith. Tiny calcium carbonate platelet from the outer covering of certain marine algae. Typically they are only a few thousandths of a millimetre in size. Their beautiful shapes and symmetry have been properly appreciated only since the widespread availability of the scanning electron microscope.

Continental shelf. That part of the sea floor surrounding a land mass where the water is less than 200 metres deep.

Cordaites. Informal name for an extinct order of gymnosperms (one of the two main divisions of the seed-bearing plants) that dominated land habitats in the late Palaeozoic. Trees could be up to 30 metres high. Commonly preserved in coal deposits of the Carboniferous age.

Correlation. The recognition (on the basis of fossils, rock type, chemistry or magnetism) of rocks that are the same age.

Cyanobacteria. Formal name for blue-green algae.

Cycad. Order of Gymnosperms (see Cordaites) of palm like appearance. Common in the Mesozoic but gradually out-competed by modern (Angiosperm) trees in the late Cretaceous.

Daughter product. Isotope or element formed by radioactive decay.

Dendroid. Type of graptolite (thought to be ancestral to the rest of the group) that lived on the sea floor (i.e. benthic).

Diagenesis. The process by which original signatures (i.e. of isotopes) are overwritten by post-depositional chemical changes.

Diversity. Number of species or other taxonomic groups.

Dolomite. Form of carbonate with high proportion of magnesium.

Element. Fundamental unit of the chemical world with specific properties, e.g. oxygen, carbon, uranium. Cannot be broken down further by chemical means.

Epoch. Time equivalent of the rock unit known as a 'series'. Several epochs together form a 'period' and several periods form an 'era'.

Eukaryote. Organism with a system of membranes inside the cell and in which the genetic material is contained within a nucleus.

Facies. The sum of the features of a sedimentary rock, including mineral and fossil content, sedimentary structures and characteristics of bedding. Facies analysis is often useful in working out the original environment of deposition.

Foraminiferan (foram). Single-celled creature that secretes a hard shell of calcium carbonate. May be either free-floating in the ocean (planktonic) or living on the sea-bed (benthic). The shells (or tests) of these creatures are very widely used in palaeoceanographic and palaeoclimatic studies. Planktonic foraminifera are used to monitor surface and near-surface conditions (particularly of temperature and productivity using oxygen and carbon isotopes respectively) while benthics are used to monitor deep ocean conditions. The contrast between the two is often vitally important in calculating temperature gradients and the rate of carbon dioxide change in the geological past.

Genome. The total genetic complement of an organism.

Genotype. The specific genetic 'blueprint' of an organism.

Geochronologist. A scientist who specialises in determining absolute (as opposed to relative, see Biostratigraphy) ages for rocks, fossils and minerals.

Gingko. A type of Gymnosperm common in the Jurassic but now represented by only one species, *Ginkgo biloba* or the 'maidenhair tree'.

Goniatite. A sub-class of the ammonoids characterised by relatively simple shell ornamentation.

Graptolite. Informal name for a class of extinct marine animals that lived together predominantly in giant floating colonies and which secreted a hard external skeleton of protein. Each individual polyp in the colony lived in its own tube (or theca). These tubes were arranged into strands (stipes) which could be single or double rowed, curved or straight, branching or not, depending on the species. When found in rock these fossils often have a very distinctive 'hacksaw-blade' aspect. Graptolites are the most widely used fossil for the biostratigraphic correlation of the Palaeozoic era. They became extinct near the beginning of the Carboniferous period.

Graywacke. Sandstones that contain a high proportion of clay.

Greisbachian. Stage near the base of the Triassic system.

Halothermal circulation. The type of deep water circulation that is driven by the sinking of salty waters in the low latitudes. Evaporation increases the density of this water (with increased saltiness as a necessary corollary) which sinks as a result and flows out into the world's oceans. The ancestral ocean of the present-day Mediterranean (Tethys) is likely to have exhibited this rare form of deep water circulation. Contrast with Thermohaline circulation (see below).

High latitudes. Near the poles.

Illite. A type of clay.

Inflection. Commonly used term when describing sudden changes in a stable isotope record (e.g. of oxygen or carbon in a deep-sea core). 'Excursion' is another term.

Isotopes. Varieties of elements that have identical chemical properties (as determined by the number of electrons and protons) but varying atomic mass. Unstable isotopes spontaneously change into other isotopes and/or elements and the time taken for this to happen is statistically a constant. Hence they form the basis for the science of geochronology. Stable isotopes do not spontaneously change into other isotopes or elements but their mass differences often result in their separation during physical or chemical processes (i.e. temperature change or photosynthesis). Together with their long life-span this means that they can be used to trace various physical processes over geological time.

Kaolinite. A type of clay.

K–T boundary. Common abbreviation for the Cretaceous–Tertiary era boundary when the dinosaurs became extinct. Note that 'C' is not used because of other geological periods that begin with this letter. The derivation of the 'K' is almost certainly from the German for Chalk which is 'Kreide'.

Lagerstatten. Fossil-bearing strata where the quality of fossil preservation is exceptional.

Lineage. A succession of related species. Often associated with the concept of phyletic speciation which is now thought to be uncommon in the fossil record.

Low latitudes. The tropics.

Macroevolution. The results of evolution as seen in the rock record.

Mass spectrometer. Instrument for separating isotopes on the basis of their relative mass.

Meiosis. The division of a sexually reproducing organism's genome into two halves is a necessary prerequisite for sexual reproduction if doubling of the quantity of genetic material is to be avoided at each generation. Meiosis (or 'reduction division') is the name given to this process. Meiosis and the merging of different genetic materials at fertilisation generate variability for evolution to work on.

Mesophytic. Assemblage of land plants characteristic of the Mesozoic era.

Microevolution. The tiny changes that occur in the genotype as a result of meiosis or mutation. Bridging the conceptual gap between microevolution and macroevolution was the primary task that the 'Trinity' of Simpson, Mayr and Dobzhansky set themselves in the forties and fifties.

Milankovitch hypothesis. The idea that the succession of glacial ages in the Pleistocene is dominated by changes in the three components of Earth's orbit (eccentricity, obliquity and precession).

Monograptid. A type of graptolite characterised by a single straight stipe with only one row of thecae.

Morphology. The exterior shape of an organism.

Morphotype. The general form of a morphology particularly when comparing groups of organisms. For example, planktonic foraminifera exist as two generalised morphotypes, Globigerine (globular) and Globorotaliid (flattened globular).

Norian. A stage of the Triassic.

Order of magnitude. Ten times the previous number (i.e. 100 is one order of magnitude greater than 10).

Organelle. A component of the cell that carries out a specific function.

Orthocone. A type of cephalopod mollusc with a tapering calcareous shell common in Palaeozoic seas.

Outcrop. Rock exposure on land.

Palaeoceanography. The study of ancient oceans.

Palaeocene. The first series (or period) of the Cenozoic.

Palaeoclimatology. The study of ancient climates.

Palaeogene. The first half of the Cenozoic era, encompassing the Palaeocene, Eocene and Oligocene periods and the interval from 65 to 25 million years before present.

Palaeophytic. The assemblage of plants that dominated the land before the Permo–Triassic boundary.

Palaeozoic. The first era of the Phanerozoic eon (from approximately 570 to 220 million years before present).

Pangea. The single supercontinent that dominated Earth from the late Permian to the end of the Triassic. The universal ocean that surrounded it is known as Panthalassa. The formation of Pangea is thought by some to be the cause of the Permo–Triassic extinctions as shallow-ocean habitable area was reduced.

Phanerozoic. The eon that encompasses the era of multicellular life which left fossilised (mostly calcium carbonate) hard parts (540 million years ago to the present day).

Phenotype. The total physical appearance and functioning of an organism (shape, physiology, etc.). The exterior expression of the genotype.

Phenotypic variability. The ease with which a group shows variation of form (i.e. size or shape) through time.

Phyletic speciation. The idea that one species will give rise to another single species by gradual transformation through time. A succession of such species is called a 'lineage'.

Phylogenetic. Pertaining to evolution.

Pleistocene. The most recent 2.5 million years of Earth's history. Best known for its succession of glaciations and the fact that mankind attained sentience during its tenure.

Precambrian. Common term for the two eons that precede the Phanerozoic eon, the Archaean and the Proterozoic.

Productivity. The rate at which organic carbon is produced (by photosynthesis) in the surface waters of the ocean.

Prokaryote. Organism which lacks membranes within its cells. Consequently the genetic material of prokaryotes is not contained within a nucleus. Includes many bacteria and the blue-green algae.

Proterozoic. The youngest of the two main divisions of the Precambrian (the other is the Archaean) extending from the top of the Archaean 2.5 billion years ago to the base of the Phanerozoic 540 million years ago. Commonly subdivided in ascending order into the Palaeoproterozoic (2.5 to 1.6 billion years ago), Mesoproterozoic (1.6 billion to 900 million years ago) and Neoproterozoic (900 million to 540 million years ago). Proterozoic rocks contain chemical traces of life (chemofossils) as well as the physical remains of bacteria and blue-green algae.

Pteridophyte. Flowerless green plants which include the ferns, horsetails and club-mosses. They evolved during the Silurian but gradually became less important as the seed-bearing plants came to dominate Earth.

Pteridosperm. Extinct order of Gymnosperms which were the earliest seed-bearing plants. They dominated the world during the Carboniferous when most of the coal measures were laid down but died out in the Cretaceous.

Radiogenic. Isotope or element formed by the decay of a radioactive parent isotope.

Radiometric dating. Numerical age estimates made by measuring the ratio of a daughter-isotope to its parent e.g. the decay of uranium to lead.

Respiration. The process of extracting energy within the cell from food.

Rhaetic. Informal name for the Rhaetian, a stage of the Upper Triassic.

Scanning electron microscope (SEM). A microscope that uses a beam of electrons to scan the surface of an object and generates an image using the intensity of the reflected electrons. Magnifications of up to 100,000 times are possible. The short wavelength of electrons means that you can view the three-dimensional topography of tiny objects such as planktonic foraminifera and nannofossils in exquisite detail.

Section. A geological outcrop on land.

Seismogram. Diagram showing the nature and ordering of rock strata generated by exploiting the different length of time required for sound to travel through various rock types.

Semantide. Complex organic molecule that retains a record of its own evolutionary history. The ultimate (or primary) semantide is DNA itself, but the products of DNA (RNA and proteins) also preserve evolutionary information and, prior to the advent of DNA sequencing technology, were in practice easier to measure.

Sequencing. The working-out of the order of nucleotide bases along a strand of DNA or RNA.

Series. The rock equivalent to the geological time unit known as an 'epoch'. For example all of the rocks laid down in the Wenlock epoch comprise one series. Usually there are several 'stages' (rock units representing a smaller unit of time) within a series and several series make up a larger unit known as a 'system'. A 'system' is the rock equivalent of the time unit known as a 'period'.

Spectral analysis. The mathematical technique by which long time-series data (i.e. oxygen isotopes down a long deep sea core) can be summarised to show the dominant frequencies of change. To illustrate by example, a typical core of Pleistocene age shows a mishmash of wiggles that look incomprehensible. Spectral analysis demonstrates that in fact the wiggles fall into three distinct categories: change with a period of 100,000 years, 41,000 years and 23,000 years. These correspond to the so-called Milankovitch parameters, changes in the eccentricity of the Earth's orbit, the obliquity of the plane of the ecliptic, and the precession of the equinoxes respectively.

Stage. See Series.

Stratigraphy. The study of sedimentary rocks, especially their ordering in time through the use of fossils, rock type, or chemical or magnetic changes (the techniques of bio-, litho-, chemo- and magneto-stratigraphy respectively).

Subduction. The process by which oceanic crust is consumed at ocean margins (in subduction zones) by sliding beneath continental plates and thence back into the interior of Earth.

Superposition. (Law of) Younger sedimentary rocks are always deposited on top of older rocks even if subsequent Earth movements change this relative ordering.

System. See Series.

Systematics. The name given to the science of the classification of organisms.

Taxon. Loose term for one of several hierarchical categories into which organisms are catalogued. A species (e.g. *Homo sapiens*) is a high-level taxonomic category (taxon), a phylum (e.g. the mammals) a lower-level one.

Tethys. Circum-equatorial ocean that existed for most of the Mesozoic era separating two supercontinents, Laurasia to the north and Gondwanaland to the south. Important in palaeoceanographic studies as the possible site of Warm Saline Deep Water formation that resulted in a reversed circulation in the world's oceans. Tethys closed in the early Cenozoic. The present-day Mediterranean is only a remnant of this once enormous Earth-girdling sea.

Tetrapod. The vertebrates that possess four limbs, the amphibians, the reptiles, the birds and the mammals.

Therapsid. A member of a group of reptiles now known to be ancestral to the mammals. Both share the fact that they have only one temporal opening in the skull. The Therapsids were the dominant land vertebrates from the late Permian to the early Jurassic.

Thermohaline circulation. The type of deep ocean circulation that operates today. Deep waters are induced to sink by cooling in the high latitudes (the Norwegian and Weddell Seas) from where they flow away under the influence of Earth's rotation until they arrive in the north Pacific approximately a thousand years later. As they circulate they acquire more of the light isotope of carbon from the 'plankton rain'. Therefore the amount of light carbon in the water (and the fossils making their shells in it) is proportional to the time since that water left its source area. Comparison of carbon isotope measurements from different localities of about the same depth therefore results in the ability to map deep water circulation pathways (compare with halothermal circulation).

Transcription. The synthesis of a strand of RNA by matching nucleotides with a DNA template.

Trilobite. A marine arthropod (i.e. related to crabs and insects) common in the Palaeozoic. This group died out during the mass extinction at the Permo-Triassic boundary.

Zonation. A scheme for the division and recognition of small intervals of geological time using fossils (see zone).

Zone. A unit of geological time defined by the occurrence of a specific fossil or suite of fossils. In the Cenozoic the planktonic foraminifera and the calcareous nannofossils are the groups most widely used for the zonation of marine sediments, in the Mesozoic this role is taken by the ammonites, and in the Palaeozoic by the graptolites.

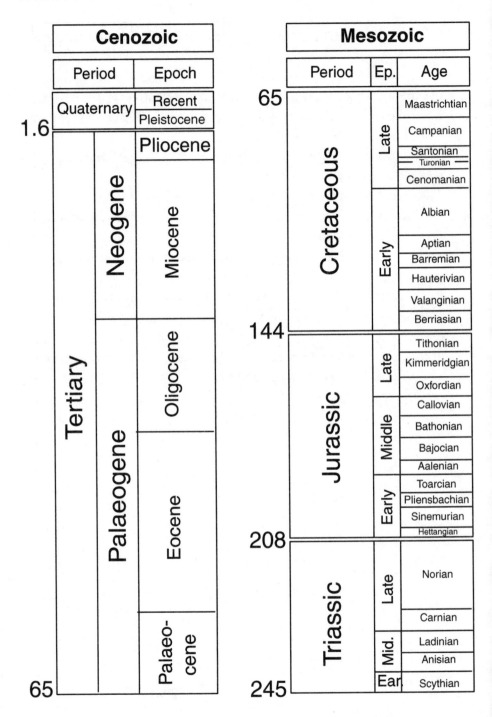

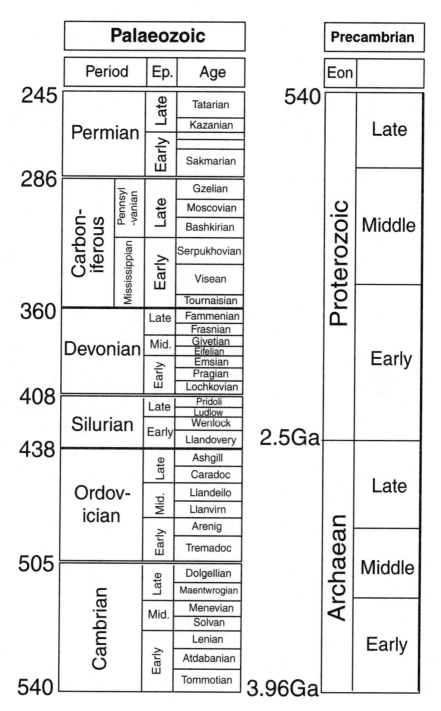

Palaeozoic				Precambrian	
Period	Ep.	Age		Eon	
245				**540**	
Permian	Late	Tatarian		Proterozoic	Late
		Kazanian			
	Early				Middle
		Sakmarian			
286					
Carbon-iferous	Pennsyl-vanian / Late	Gzelian			Early
		Moscovian			
		Bashkirian			
	Mississippian / Early	Serpukhovian			
		Visean		**2.5Ga**	
		Tournaisian			Late
360					
Devonian	Late	Fammenian		Archaean	
		Frasnian			
	Mid.	Givetian			
		Eifelian			Middle
	Early	Emsian			
		Pragian			
		Lochkovian			Early
408					
Silurian	Late	Pridoli			
		Ludlow			
	Early	Wenlock			
		Llandovery			
438					
Ordov-ician	Late	Ashgill			
		Caradoc			
	Mid.	Llandeilo			
		Llanvirn			
	Early	Arenig			
		Tremadoc			
505					
Cambrian	Late	Dolgellian			
		Maentwrogian			
	Mid.	Menevian			
		Solvan			
	Early	Lenian			
		Atdabanian			
540		Tommotian		**3.96Ga**	

Selected Reading

Chapter 1. The Size of Things to Come
Desmond, A. *Huxley: The Devil's Disciple* (Michael Joseph, 1994)
Schuchert, C. *O.C. Marsh, Pioneer in Paleontology* (Yale University Press, 1940)
Lageson, D. and Spearing, D. *Roadside Geology of Wyoming* (Mountain Press Publishing Company, 1988)
Lanham, U. *The Bone Hunters* (Columbia University Press, 1973)
McPhee, J. *Rising from the Plains* (Farrar, Straus and Giroux, 1986)

Chapter 2. The Hunt for the Ruler of Time
Arkell, W. J. *The Jurassic Geology of Great Britain* (Oxford University Press, 1947)
Elles, G. L. 'The zonal classification of the Wenlock Shales of the Welsh borderland' (Quarterly Journal of the Geological Society of London, 1900)
Siveter, D. J., Owens, R. M. and Thomas, A. T. *Silurian Field Excursions: A geotraverse across Wales and the Welsh borderland* (National Museum of Wales and The Geologist's Association, 1989)

Chapter 3. The Science of Real-time
Pledge, H. T. *Science since 1500; A short history of mathematics, physics, chemistry, biology* (HMSO, 1966)

Harper, C. T. *Geochronology: Radiometric dating of rocks and minerals* (Benchmark papers in Geology series, Dowden, Hutchinson and Ross, 1973)

Kennett, J. *Magnetic Stratigraphy of Sediments* (Benchmark papers in Geology series, Dowden, Hutchinson and Ross, 1980)

Lewis, C. *The Dating Game* (Cambridge University Press, 2000)

Snow, C. P. *The Physicists: A generation that changed the world* (Macmillan, 1981)

Chapter 4. Weighing Oceans

Bowen, R. *Paleotemperature Analysis* (Elsevier, 1966)

Emiliani, C. 'Ancient Temperatures' *Scientific American* (February, 1958, p. 54–63)

Norris, R. and Corfield, R. M. *Isotope Paleobiology and Paleoecology* (Paleontological Society, 1998)

Chapter 5. The Savage Hand of Evolution

Gee, H. *Deep Time* (Fourth Estate, 2000)

Schopf, T. *Models in Paleobiology* (Freeman, Cooper, 1972)

Simpson, G. G. *The Meaning of Evolution* (Yale University Press, 1949)

Simpson, G. G. *This View of Life* (Harcourt, Brace and World, Inc 1947)

Williamson, P. *Palaeontological documentation of speciation of Cenozoic molluscs from the Turkana basin* (Nature, 1981)

Wyndham, J. *The Day of the Triffids* (Michael Joseph, 1951)

Chapter 6. Heavy Metal and the Italian Rock Band

Alvarez, W. *T. Rex and the Crater of Doom* (Penguin, 1998)

Frankel, C. *The End of the Dinosaurs* (Cambridge University Press, 1999)

Raup, D. *The Nemesis Affair: A story of the death of dinosaurs and the ways of science* (Norton, 1986)

Cresta, S., Monechi, S., and Parisi, G. *Mesozoic-Cenozoic stratigraphy in the Umbria-Marche area: Geological Fieldtrips in the Umbria-Marche Apennines (Italy)* (Istituto Poligrafico e Zecca dello Stato, 1989)

Chapter 7. The Terminators

Bains, S., Corfield, R. M., and Norris R. D. 'Mechanisms of climate warming at the end of the Paleocene' *Science* (July 1999)

Schlanger, S., Arthur, M., Jenkyns, H, et al. 'Stratigraphic and paleo-oceanographic setting of organic carbon-rich strata deposited during Cenomanian-Turonian oceanic anoxic event' *American Association of Petroleum Geologists Bulletin* (1983)

Schmitz, B., PeuckerEhrenbrink, E., Lindstrom, M., et al. 'Accretion rates of meteorites and cosmic dust in the Early Ordovician' *Science* (October 1997)

Walliser, O.H. (ed.) *Global Events and Event Stratigraphy in the Phanerozoic* (Springer-Verlag, 1996)

Wignall P. and Hallam, A. 'Anoxia as a cause of the Permian-Triassic mass extinction' *Palaeogeography, Palaeoclimatology, Palaeoecology* (May 1992)

Chapter 8. Beyond the Kingdom of the Small

Lewin, R. *Patterns in Evolution: The new molecular view* (Scientific American Library, 1996)

Zuckerkandl, E. 'The Evolution of Hemoglobin' *Scientific American* (May 1965)

Doolittle, W. F. 'Uprooting the Tree of Life' *Scientific American* (February, 2000)

Chapter 9. The Organ Grinders

DeSalle, R and Lindley, D. *The science of Jurassic Park and The Lost World* (HarperCollins, 1997)

Mullis, K. *Dancing Naked in the Mind Field* (Bloomsbury, 1998)

Wayne RK, Leonard JA, Cooper A. *Full of sound and fury: The recent history of ancient DNA* (Annual Review of Ecology and Systematics, 1999)

Chapter 10. Battles of Deepest Time and Space

Conway Morris, S. *The Crucible of Creation* (Oxford University Press, 1998)

Gould, S. J. *Wonderful Life* (Hutchinson Radius, 1989)

Hoffman, P and Schrag, D. 'Snowball Earth' *Scientific American* (January, 2000)

Schopf, J. William *Cradle of Life* (Princeton University Press, 1999)

Index

'f' indicates a figure and 't' indicates a table.